"Famulari makes a public call to the design profession in her new book to integrate greening into everyday practice. Each individual involved in shaping the built and natural surroundings has the responsibility to lean on allied professions to create healthier places for people and this book is the recipe you'll want to follow. Inequitable access to green space was exacerbated by the pandemic. It's our responsibility to right these inequities and Famulari puts designers, planners, architects, landscape architects, and other allied professions on notice to step into the role of a change agent. Professor Famulari presents her most important lecture yet. The teachings in this book highlight the vision to easily integrate greening practice.

"Whether you are new to this work, or a veteran, this book provides green selections of design interventions that can serve as inspiration for a range of contexts and applications. The time is now for greening and biophilic design – health cannot wait one more day. And Famulari shows us just how easy it can be."

– **Ashley Zidon, ASLA, AICP,** *Community planner, researcher, and change-maker*

"This book addresses the impacts on design and the need to build an inclusive culture – detailing the how and who has access to interact with green space. Stevie's ability to challenge the status quo, specifically with how people and plants interact, is a beautiful push on our responsibility to climate change. She does an excellent job showing us how to prioritize and value plant life and its numerous positive impacts on the earth and life in every facet of daily living."

– **Janna Pea,** *Founder of Pea Nation, communications executive specializing in social impact campaigns and narrative change*

"As a traveler of the world I have seen how different societies use green design in a variety of spaces. Through an ecological lens, I've encountered communities with wonderful unique approaches. Stevie Famulari Gds integrates best practices from different cultures into her green designs in order to create new Ways of Greening. These beautiful renditions effortlessly weave issues of DEI, health, joy, urban air quality, and shared public spaces into fresh approaches for Greening Design. This third book of Stevie Famulari, Gds, continues to expand boldly and creatively on her unique ways of Engaging Green."

– **Henry Obispo,** *Eco-Gastronomer & Founder of ReBORN FARMS*

Ways of Greening

This book focuses on rethinking working and living spaces and understanding how "greening" can make them healthier and their occupants happier. It teaches how to see unique ideas for spaces and some of the materials needed to create the designs.

Inspired by a study that states that 8% of a space needs to have plants in order to positively affect the air quality of the space, this book explores what that minimum would look like in spaces and how it can be done to existing spaces as well as to new site design, greening both interiors and exteriors. Using the mathematical amount of 10% per square foot, the illustrations start at that quantity of greening and show how it can look.

The sites selected are both public and private sites, as well as interior and exterior. As there are more modalities, needs, and locations where people now work, making sure that multiple types of spaces are designed for people's success is more relevant than ever. This includes designs for more traditional offices, open-air offices, commercial spaces, homes, studios, and more.

Ways of Greening: Using Plans and Gardens for Healthy Work and Living Surroundings gives readers a way to not only understand greening but to understand how to see greening applied to their place. The two basic ways to see the spaces selected are existing spaces to which greening design is applied afterward and upcoming spaces in which greening design can be built directly into the space. The first type of retrofitting greening into existing spaces can also be combined with the second type of space (new designs). There are examples of both types throughout the book. Essentially, this book addresses ways in which business owners, residents, developers, architects, agencies, and others can integrate greening to improve the air quality and the quality of life with a green solution.

Ways of Greening

Using Plants and Gardens for Healthy Work and Living Surroundings

Stevie Famulari, Gds

Routledge
Taylor & Francis Group

A PRODUCTIVITY PRESS BOOK

First published 2024
by Routledge
605 Third Avenue, New York, NY 10158

and by Routledge
4 Park Square, Milton Park, Abingdon, Oxon, OX14 4RN
Routledge is an imprint of the Taylor & Francis Group, an informa business

ISBN: 978-1-032-39155-7 (hbk)
ISBN: 978-1-032-39154-0 (pbk)
ISBN: 978-1-003-34863-4 (ebk)

DOI: 10.4324/9781003348634

Typeset in Frutiger LT Std
by Deanta Global Publishing Services, Chennai, India

*To creative people who move forward with amazing ideas, bringing
life and greening for today and the breathtaking future.*

*To Shelly and Missy – you are my magic, my life, my laughter, my love, my forever.
And you both are why I am always thankful for my fantastically amazing life. Forever.*

Contents

Foreword by
Kene Okigbo, PLA, ASLA

The following pages of this book act as a continuation of a conversation. It's fine for you to jump in now, or it may help you to be familiar with the premises of the first two books. Book one, *Green Up! Sustainable Design Solutions for Healthier Work and Living Environments*, is an introduction to the concept that increasing the live vegetative massing in and around our built world has positive wellness impacts on people. This concept will hereon be referred to as "greening." Book two, *Designing Green Spaces for Health: Using Plants to Reduce the Spread of Airborne Viruses*, expounds on the same principles of greening but does so by applying them via the *Famulari Theory*. This theory posits that with six specific preconditions met, positive wellness impacts may be achieved as the propensity for infection via the influenza virus is reduced. Book three imagines what existing and proposed spaces can become if greening principles are embedded in and around them.

What is most fascinating about this trilogy is the subtle subterfuge that takes place without the reader noticing. I would call greening an example of revisionist futuring. The ideas and illustrations that occupy the coming pages are glimpses into the mind of Stevie Famulari. They are an idealistic expression of what the world either could or should be. Famulari takes multiple examples of spaces that exist in the world today and asks the reader to look at them with eyes anew. Each of those existing spaces is then re-imagined in a style that can only be described as "Famularized." The aesthetic is comfortably solarpunk, in that it represents a world that actualizes a sustainable future, built through the interplay of nature and community. Buildings are clad in vegetation in a way that highlights the functional contribution of the plant matter. Sustainable technologies are utilized to reject climate change, not as a matter of fiction, but rather to reject it as an inevitability. These technologies are represented as being utilized intentionally and intelligently. They are specifically employed to increase the well-being of those that occupy these spaces.

The solarpunk philosophies, though not referred to as such in this book, are visionary in the sense that they stand firmly in Famulari's experience and research in phytoremediation and other effects that plant matter can have on their surroundings. This book is a reimagining of a future state where our technology is ethically advanced and maximizes the opportunities presented through nature.

What is most powerful about this book is its view of nature as a technology to be integrated with the built milieu. Through this philosophy our technology is phytotechnical and relies on the therapeutic, physiological, and phytoremedic properties of various plant species to influence anthropological and ecological wellness.

In *Ways of Greening*, Famulari essentially establishes new design typology under the guise of a hyper-simplified premise. The ideas and execution are complex, while the terminology is simplified, specifically, to make the principles digestible.

This is not an imagined future, but rather a revisualization of our present, and one that embraces nature as an extension of our contemporary building tools. We need to imagine what our world can and should be. Dreaming with open eyes is the requisite step before challenging norms, which is what this trilogy successfully does.

Kene Okigbo, PLA, ASLA

About the Author

Stevie Famulari, Gds, is an artist, author, researcher, greening design specialist, founder and principal of *Engaging Green* and professor of Landscape Architecture. She has been a professor at the University of New Mexico (UNM), North Dakota State University (NDSU), SUNY Farmingdale, and the University of Cincinnati (UC). She has guest lectured at Harvard University, Northeastern University, University of Minnesota, and University of California Berkeley. She is a keynote speaker, received numerous awards, and has shared information to help numerous people experience greening in new and unique ways.

This is Stevie's third book. Her first book, *Green Up! Sustainable Design Solutions for Healthier Work and Living Environments,* explores unique greening solutions and practices that help create a lifestyle shift, improving the health of living and working spaces for its occupants from an individual, business, ecological, and profitable perspective. Her second book, *Designing Green Spaces for Health: Using Plants to Reduce the Spread of Airborne Viruses*, focuses on using plants in spatial design to reduce the infectiousness of viruses in different working and living spaces. It presents strategies for interior and exterior green designs with plants that are likely effective for flu virus tolerance and reduction of infectiousness.

Her installation *Engaging Urban Greening* was installed at the Smithsonian Museum in Washington D.C. Over 1,400 seeded paper pieces (colorful papers embedded with a wildflower seed mix) were given to the public during the installation. People from D.C., New York, Michigan, Maryland, Florida, California, Oregon, Nigeria, Melbourne Australia, India, Yorkshire England, and South Korea added green to their community by planting their seeded papers in the soil, creating their own living artwork.

Mx. Famulari's designs and design research explore the relationship, extension, and application of green designs to other diverse fields to create healthy spaces for living and working. Time, change, and engaging in the process are some of the continuing themes in her works. Defying traditions and celebrating ecological processes, using community and individual stories of sites, are themes she is widely recognized for. With the applied art of green design combined with the science of phytoremediation, her works have aesthetic beauty as well as healing properties for both people and the environs.

Her work in greening designs, research, and education can be seen nationally and internationally and includes: green walls; planted roofs; green office designs, green remediation designs for interior and exterior applications; designs for the Environmental Protection Agency in Colorado; Ghost Ranch Visitor Center for the Georgia O'Keefe Museum in Abiquiu, New Mexico; developing green design and policies on the University of New Mexico campus; flood control design for the Red River in North Dakota.

Lectures, presentations, and exhibitions of her work include Plains Art Museum, San Diego Museum, Minnesota Erosion Control Association (MECA), Engineering Mechanics Institute, University of Minnesota, (UMN), NDSU, American Society of Landscape Architecture (ASLA), and the American Institute of Architecture (AIA). Awards of Stevie Famulari's works and community leadership include the Ecological Leadership Award from MECA, Community Service Award from The Arts Partnership (TAP), selected for team USA for the Festival Du Voyageur, Art in Public Places in New Mexico, New York State Floral Industry (NYSFI), Long Island Nursery and Landscape Association (LINLA), and NMEH Award for New Mexico Symphony Orchestra.

Stevie Famulari's phytoremediation database of hundreds of plants which clean the air, soil, and water of contaminants is used by the EPA, courses at Harvard Graduate School of Design, International Phytoremediation Society, landscape architecture and engineering firms, and administrative agencies. Based on her second book, Stevie also created a database of plants for reducing airborne viruses. She has been an investigator for grants which explore water remediation design for oil drilling and fracking processes, improving air quality, remediation design for communities, and interior greening applications.Her designs and design research explore the relationship, extension, and application of green designs to other diverse fields. Stevie's PhD, ABD research is at RMIT. She received her graduate degree in Landscape Architecture (MLA) from the State University of New York (SUNY) ESF with a concentration in Fine Arts from Syracuse University. Her undergraduate degree is in Fine Arts (BFA) from NYU.

Stevie Famulari's work has appeared in hundreds of books, magazines, newspapers, and television programs, including Smithsonian Magazine, Food Network Challenges & Specials, Oakland Tribune, World Entertainment News Network, The Oprah Winfrey Show, Travel Channel, Good Morning America, CBS Early Morning News, The Washington Post, The Post-Standard, Trust for Public Land, Boston Herald, Berkeley Daily Planet, Santa Fe Reporter, and Star Tribune.

In the Land of Gifted Futures

I stand alone before the precipice of oblivion, debating over whose footprints led to this position.

Negotiating with them over crumbs in the wake of a feast already devoured.

But when I close my eyes, I dream lucidly of worlds not yet made.

I dream of streets of sand, flooded with glass people.

I dream of palms, both uplifting and life-bearing.

I dream of water being respected and deliberately upcycled.

When my eyes open, I speak through sealed lips as my hands caress pen on paper.

I speak my dreams into existence.

I see streets unsaddened; lush with life, prioritizing those who have it.

I see palms; fully sunned, outstretched in a constant attempt at embrace.

I see systems structured for the sustainability of mind, body, and soul

With this, I commit my body to the ground. I commit her from earth to earth. From concrete to compost. And from dust to dusk.

Kene Okigbo

Giving Gratitude and Thanks To

For all the people who I am honored to have as my friends, family, and colleagues:

Ashley – Continue to change, move forward, and be fantastic with all that you amazingly do

Charlie – Your combination of twisted humor and depth of caring is only found in you!

Grit City crew – You create a twisted community of support that I couldn't be prouder to be a part of

Herb – For always the toughest at any table, and redefining what tough truly means

Janna – My fearless friend who changes the world in ways even bigger than you

Kene – I look forward to *your* books soon.

Kimba – Who would I be without you as my friend. I don't want to even imagine it!

Kris – For pizza, wine, depth, laughter, and sharing your brilliant art

Lisa – For walks, salt, soaks, lavender, and light

Missy – The Goddess, you fiercely love and live and make me love bigger and infinitely better, my forever love

Nonni – Because you are fearless and courageous

Patrick – Keep building everything that has never been built

Paul & Shane – Across the world, you are still two cool people who I adore.

Shelly – The Goddess, love in the fuzziest form, love beyond love, love x love x love x infinity, my forever love

Sophie – For surfing, pizza, wine, depth, laughter, and staying left when driving

Introduction

Image i.1 Green design of downtown New York City. The original drawing is about 4 feet long and took a few years to complete. I was creating this green design during my free time between site designs, teaching courses, and writing and designing other books, seeing how much green space I could add to an expansive site.

Image i.1 is the initial greening illustration which started my thought process for this book – a series of green designs which can be used to retrofit or redesign existing spaces. From this original image the full book grew.

I have been doing these green drawings and designs for a myriad of years – small sketches, large colorful pieces, in a sketchbook, framed drawings, etc. They take minutes, hours, days, weeks, and years for the really large ones. It is not only about drawing it; sometimes it is about seeing it in my head so that I can get the images onto paper. The designs in this book and countless others are how I truly see the spaces I have been to. I see it for what it could be. This is the typical way I think.

The drawings in this book are how I see existing spaces in my thoughts with green design as I walk or drive in different locations. I see unique greening designs everywhere, horizontal, vertical, small, large, upside down, and more. I see greening designs when I walk, when I go for a run, when I am waiting for a train, when I am driving – I see green everywhere!

Image i.2   Initial sketches for a green office with a variety of living styles for different areas of the open office plan.

This way of seeing green design started several years ago. More recently I realize that count-less other brilliant people have created urban forestry in unique ways and that these designs are achievable. A few events led me to this realization. One is an email dialogue with Stefano Boeri. I had done research on his designs including the Bosco Verticale (which can be seen in Chapter 2) and other designs by him and his company. His integration of urban forestry is original and achievable.

Stefano Boeri graciously agreed to write a piece for one of my books, *Designing Green Spaces for Health: Using Plants to Reduce the Infectiousness of Airborne Viruses.*

GREEN DESIGN

One inspiration for this book is quite bizarre: the inspiration was in contrast to green design ideals.

I am in a meeting for designing a new office space in New York City, discussing an office biophilic (mimicking nature or natural processes) design with greenwalls and office gardens throughout the expansive open floor. The architects are showing the clients a design which was poor in biophilic plans, health, or even having the green components live for more than a couple of months. The architects arbitrarily put random plants into the digital images to fill in the space without any consideration to correct indoor plants, lighting, growth, and using green for design-ing. My role is to address this.

I suggest a series of living green wall art using mostly natural sunlight from the windows. However, there is one space – the entry – which has no natural sunlight nor windows and there-fore requires artificial full spectrum plant lights, likely LED lights, in order for the plants to live, let alone grow.

During one exceptional meeting, an architect says that he is not interested in putting plant lights for the plants to live and grow. He says "Aren't we really just fighting Mother Nature?!". I say

"We see the world differently. Plants, gardens, green healthy lifestyles belong in both interior and exterior spaces. Green design belongs everywhere there are living beings – people, animals, plants – they all work together when you care to. Greening Design belongs everywhere to improve both physical and psychological health. It belongs everywhere and can be designed directly into the spaces as part of the design process for both short-term and long-term design. That is more effective than an arbitrary afterthought and pretty sketch that won't live in any way or show any knowledge of green design."

I continue:

"Plants have needs just as all living things do. And we consider people's needs of design, light, water, seats, desks, heights of people, ability, etc. But you cannot have enough consideration to design for other living things such as plants?! They are living and have basic needs of light, water and space to grow. If you are creative enough for designing a space, do better and design with living green designs for everyone's benefits."

The challenging dialogue with the architect reminds me that the health of all parties – people, animals, plants – can all work together with consideration for all. And that not only can it be done successfully, it has already been done by others before me. Alan Sonfist, Mary Miss, Buster Simpson, Agnes Denes, and Meg Webster, to name a few, create living designs. And you can also join the crew of creative people to add healthy ecological design to your space.

There is one other notable inspiration, a brief dialogue I had with Kene Okigbo over a decade ago. Kene Okigbo is the author of the Foreword of this book. Years ago I had Kene as a student in my landscape architecture studio at the university I was teaching. He thought about an idea for New York City parks to not be large public space,

Form and aesthetic appeal of green design

– Designing spaces, vs. adding plants as objects- discuss the difference
– Design needs- what does one address in order to design a planted green space effectively

Additional pieces of greening design

– Organized in an aesthetically interesting design
– Colors
– Forms- heights, shapes, textures
– Sounds- wind through leaves or tall grasses, water features
– Style of navigation through the space- walkways, paths
– Views
– Changes throughout the year and years to ahead
– Scent
– Variety of plant scales, heights, colors
– Vertical and horizontal green components
– Blooming at different months of the year
– Creating magical interventions for unexpected beauty

For plants to live well

– Light- hours and distance to sunlight
– Light- hours and distance to artificial light
– Light- artificial light temperature (degrees kelvin)
– Water
– Temperature
– Nutrition
– Soil

Image i.3   List for visualizing some of the pieces and layers which go into green design.

but rather small interventions of plants embedded throughout a specific travel route. Kene then went onto a different project and continued his career as a landscape architect advocating emergent design typologies and improved equitability of landscape architectural practices.

Even years later, it is that one initial concept of smaller embedded interventions that continues to inspire me and one I have used as a starting point toward expanding greening design. In this book, parks, gardens, and green design do not have to be a destination, but rather a design choice, a design insertion, or a healthy ecological approach toward improving spaces. Green designs can be paths in urban sidewalks with living gardens, or light posts with plants, storefront awnings with living green design, and storefront windows with unique greening designs. The concept of ecological design insertions intrigues me with countless possibilities of applications.

Greening design has lists of forms and functions that need to be considered and integrated in order to build an effective living design. A partial list of forms and aesthetic pieces which are explored in green design are shown in Image i.3.

THE SCOPE OF THIS BOOK

The first chapter of this book explains terminology and gives further information on points which are illustrated in designs throughout the remaining chapters. Chapters 2 through 5 each take a distinct type of space and showcase the opportunities for available greening. Each chapter is built around key examples which are accompanied by notes about the design, approaches, and features to consider in each type of site and green design.

There are images of several sites with photographs of the same site showing how the site presently is, and green design illustrations presenting how they can be redesigned for ecological health. There is a larger illustration for details as well as a smaller labeled image for further understanding.

Developing new concepts and applications for gardens, parks, and green design is not always about finding existing spaces and dedicating it to be a large open green space. Consideration of new living concepts also means, retroactively greening existing spaces with unique site-specific solutions in order to reach the minimum 10% greening goal that is the healthy design green minimum application throughout this book.

This third book focuses on seeing different spaces for the greening they can be, as well as teaching people to understand how to see unique ideas for their own spaces, and some of the materials needed to create the designs.

The sites selected are both public and private sites, as well as interior and exterior. As there are more modalities, needs, and locations where people now work, making sure that multiple types of spaces are designed for people's success is more relevant than ever. This includes designs for more traditional offices to open air offices, commercial spaces, homes, studios, and more.

This book gives readers a way to not only understand greening but also to understand how to see greening applied to their site. The two basic ways to see the spaces selected are existing spaces which greening design is applied to afterward; and upcoming spaces for which greening design can be built. The first type of retrofitting greening into existing space can also be combined with the second type of space (new designs). There are examples of both types throughout the book.

Image i.4   Street scene with green interventions throughout the space.

A basic understanding of living green design is discussed to lead you to the next steps of the illustrative application of green design sections. It is a creative and knowledgeable process to develop a living green design which has an aesthetically pleasing form as well as addressing the functions and needs of the site and living beings who use it. The designs are appealing for people to interact with. Living designs improve air quality and promote healthier living with greening ideals with growing and changing designs over time.

Chapter 1

Understanding Green Design

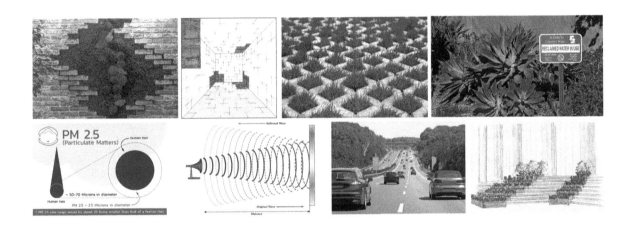

Image 1.1 Exploring greening, 10% of a space for green design, permeable pavers, grey water, air pollutants, air quality, sound waves, psychological effects of greening, and DEI.

This first chapter makes some of the intention of this book, as well as terminology, studies, and other information clear in order to have it readily accessible when reviewing the other chapters. These include:

Green, greening, green design, gardens, green spaces
10% of a space for greening
Permeable pavers
Grey water
Air pollutants and understanding air quality of a site
Sound waves
Psychological effects of greening
DEI

These practices and materials are applied to the green designs in Chapters 2 through 5. Some of the applications of these green practices and materials are more prominent in a design, while others are subtle or not included. Each of the practices and materials shows up differently, based

Image 1.2   Sites where the plants are treated as objects or pieces of the space. The plants here vary in size, yet all have the distinction that they are not integral to the space, define the space, nor respond to the conditions and structures of the space.

on the site conditions. Making sure the site's overall design addresses integration with people, plants, and animals and a healthy relationship to the site can co-exist and thrive is the balance of each green design.

PLANTS, GARDENS, GREENING SPACES OR GREEN SPACES

Defining a few terms will help to set the tone for understanding how you can use this book to inspire healthy ecological designs for the spaces you create.

PLANTS

Plants are living materials which grow in pots or planters, in the ground, or on countless vertical surfaces. They can be grown in air (such as air plants), soil, or hydroponically (in water with nutrients added to the water).

When treated as objects, plants can be moved from one place to another and are treated similarly to lamps, couches, or chairs in the sense that they are single objects and not designed into or for the space. Though the plants have different needs than the couch, such as appropriate lights, water, and nutrition, the plant is treated as an object which does not define the space but is used as a piece of the space.

GARDENS

Gardens are themed spaces in which plants and other related objects (such as seats, water features, lights, paths, etc.) are used to define the role of the space. Gardens can be themed such as vegetable gardens, annual gardens, moon gardens, children's gardens, hospital gardens, therapy gardens, grey gardens, potted gardens, and countless more.

Gardens are horizontal, vertical, or both; interior, exterior, or both; have varying amounts of living plants; and greatly vary in size and location. It is not the scale of garden that defines it as a garden, vs. plantings as objects. It is the intention of the garden as a space or spatial design, as opposed to a plant as an object which can be moved to appeal to an aesthetic in a larger space.

A garden is different than a park, whether the park is private or public. Whereas a garden typically utilizes plants, a park typically includes an activity within a green space such as baseball, running, or other sports, concerts, seating, activities for mixed ages, and more.

Image 1.3 In these sites, unique greening design in part defines the site as well as responds to the conditions of the site.

GREENING SPACES, GREEN SPACES, GREENING DESIGN, GREEN DESIGN ARE USED INTERCHANGEABLY

Greening spaces, green spaces, greening design, green design are all used interchangeably throughout this book and have the same meaning. Greening can also be used as a verb, as in the use of living plant materials for a design.

Greening design is even more expansive than a garden or park. Greening design explores using living plants and other related materials (soil, water, air circulation, etc.) for spatial designs in every scale and location for healthy living through ecological design. The idea of creating spaces which have living, growing, changing designs which form spaces of magic – no matter the location – is not only possible, it is also one of the greatest types of magic which we can do to any space.

Greening design includes gardens as well as a myriad of other green spaces such as parks, arboretums, green roofs, natural playgrounds, estate designs, living designs along traffic routes, which reduce the sound of vehicles and air pollution, tree wells, expanded tree wells to include seating and more layers of plants, and even more greening designs are being added as people become more and more creative.

Green design is where the plant or plants define the space as a notable part of it – AND the plants respond directly to the space.

A green space is a space designed with plants where plants are a notable part of the space, defining it in some way in scale or effect AND the plants are allowed to respond to the light and structure of the space as they grow. Those two conditions are the basis for the greening design.

This does not mean there needs to be a lot of plants – only that they define the space and are visually notably parts of the space. The size of the space is also not important for effective green design. The space can be large or small, and again, it is that the living plant designs define the space as well as respond to the space. The living materials are designed into the space as an integral part of the space. In greening designs, plants are not used as merely objects but designed as a spatial feature. The living green design may grow around a structure or multiple structures of the site. Structures and designs can be built specifically for the plants or not.

The idea of good and bad design is a different topic. Good and bad, or better yet – effective and ineffective, or pleasing and not pleasing design includes how people and animals can engage with the space and use or enjoy it. This includes the ease of use, views when moving throughout the space, access, lighting, colors, scents, sounds, seating, and much more. There are multiple factors which make effective and enjoyable design versus challenging and unpleasant design. The addition of greening design as an ecological approach to designing spaces typically supports enjoyable designs.

Image 1.4   Unique approaches to ecological design can be site-specific such as these sites with approaches and designs which respond to the size, materials, light, use, and structures of the space.

10% OF THE SPACE FOR GREENING

The premise of the green design, mathematically, may be seemingly simple – adding 10% more living greening design to all spaces. Yes, that is the mathematical minimum goal of the quantity of green design.

The fun challenges become HOW to do this in order to use and enjoy the benefits of the green design in the space with people, animals, and plants. The fun challenges also include being able to SEE the new design in your mind and allowing yourself to explore it before it is built.

There are a few things to consider:

First – outdoor spaces or air are not in the studies of the quantity of plants. The focus is indoor air for the quantity of plants.

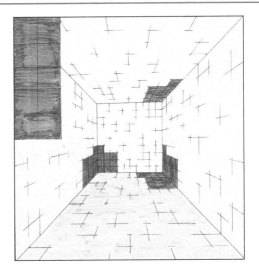

Image 1.5   Ten percent of each of the six sides of the 10-foot by 10-foot room are dedicated to greening.

Second – why 10% of the surface space? And is it 10% per square foot looking only at the ground surface?

A study in 1989 by B.C. Wolverton and others from NASA shows that 1 plant per 100 square feet (9.2 square meters) helps improve air quality. In a home of 1800 square feet (167 square meters), 15–18 house plants are needed.

Leafier bigger plants are more effective for improving air quality than other plants.

Studies since 1989 vary in defining the quantity of plants needed in a space to improve air quality, with one in 2019 (Cummings & Warren, 2020) stating that one would need 10 plants per square foot in a home to have an impact on indoor air quality. To expand that, in a room of 320 square feet, one would need 3,200 plants for the room to improve air quality.

Image 1.6   Relocating 10% of each of the green spaces of the walls, floor, and ceiling to a large green wall and floor area can be a better design solution for the space.

Other research falls between these studies in the quantity of plants and types of plants needed for improving the air quality. The specific plants, plant size, plant locations, sunlight, and water for the plant growth were not fully clear in the study for their applications to improving air quality. Nor was the size of the plant and leaves, or leaf quantity, clearly explained. Additionally, the location of the study is a lab with specific conditions, not a living or working space where most people are.

With those acknowledged, a few other items are to be considered for the 10% dialogue used as the goal for these designs. One item is square feet vs. cubic feet. Square feet is flat space, with the floor or ground plane solely used. Cubic feet uses the air in the room with the height of the ceiling and walls. This is more accurate – though more complex to design for in interior spaces, let alone the infinite sky of exterior spaces.

Image 1.7   Interior greening with free standing plant wall dividing spaces in the open loft.

In interior spaces it is more achievable to explore cubic feet. However, for the ease of these green designs, linear feet is used. Linear feet is more approachable, measurable, visible, and able to be designed for in both interior and exterior spaces. A sample of how this can be seen in a room that is 10 feet by 10 feet by 10 feet can be seen in Images 1.5 and 1.6. In exterior spaces, the ceiling of the sky is not counted toward the 10% greening.

A Bit More about Understanding the 10% Greening

First, 10% is mathematically easier to approach in design than 8%. Therefore, a starting point of greening 10% is used in designs. The goal is not always achieved, but when it falls under 10% greening, it still has a beneficial effect on the air and people who inhabit the space.

Secondly, for interior spaces, the goal is 10% of each horizontal and vertical surface. This means 10% of the total floor space, 10% of each of the four or more walls, and 10% of the ceiling is to have plants. If plants cannot be placed on the ceiling or any specific wall, the plants required to meet the 10% of the surface can be placed in other areas of the room. It is the totality of the quantity of plants that is more important than putting 10% on each of the horizontal and vertical surfaces. Images 1.7 and 1.8 visually illustrate this.

Thirdly, the size of the plants varies in scale and plant types in the spaces. For example, an entire succulent garden with small plants is not as beneficial for air quality as other plants. Using leafier water-loving plants is more beneficial for air quality as there is more leaf surface for filtering contaminants and producing oxygen. Though succulents may be part of the design, faster-growing leafier plants that will mature and fill in the space are typically more beneficial for improving air quality.

Finally, the exposure to plants in a space reduces stress, improves concentration, and helps to reduce depression and anxiety in numerous studies. Whether the green design is 5% of the space, 10%, or more, the benefits of air quality to emotional health are both relevant for the designs throughout this book.

Image 1.8   In this window garden, the plants appear to grow through and around each other. This creates interesting shapes and forms which are delightful and ever-changing for the people in the space.

PERMEABLE PAVERS

Permeable pavers have spaces for water to flow through the stones. Water on the pavers goes back into the ground and water table within the site or is used for growing plants between the pavers. In these spaces low growing plant ground covers such as grass and moss thrive. There are a variety of styles of these pavers which can be bought commercially such as those in Image 1.9, or people can create their own unique designs.

Image 1.9   Examples of permeable pavers, pavers with spaces where plants can grow, in a variety of styles and patterns.

GREY WATER

The terms "grey water" and "grey-water" are used interchangeably. This water can be used and is currently used for irrigation of planted areas. Grey water comes from sources with mild impurities and includes water used in sinks, showers, and washing machines. Mild soaps in water are not harmful to plants in small quantities. Toilets are not included in this category as their contaminants are titled "black water" with more pollutants and which damage plants.

Grey water can be stored in units and then filtered through commercial filters as needed for use. Commercial filters can be used in non-winter

FOR WATER CONSERVATION
THIS PROPERTY IS IRRIGATED
WITH RECLAIMED WATER
DO NOT   DRINK
COMO PARTE DE LOS ESFUERZOS DE
CONSERVACION DE LA CIUDAD; ESTA AGUA
HA SIDO TRATADA CON EL PROPOSITO DE
SER UTILIZADA UNICAMENTE PARA RIEGO

Image 1.10   Signage showing the use of reclaimed water used for irrigation of a site.

temperatures to remove larger particles or contaminants before the water is used for irrigation. The filters help the irrigation tubes to not be blocked with larger particles which may be in the greywater.

Grey water systems and filters are cleaned out and flushed of water before winter and restarted in the warmer temperatures of spring. During freezing temperatures outdoor water systems cannot be used so that water does not freeze within the system and damage it.

The re-use of grey water for irrigation, rather than fresh water, is beneficial when considering the larger benefit of sustainable practices.

Image 1.11   The use of reclaimed water for irrigation conserves fresh water for necessary uses such as drinking. The practice of grey water irrigation is beneficial for sustainable practices.

AIR POLLUTANTS AND UNDERSTANDING AIR QUALITY OF THE SITE

(Famulari, 2020)

The Environmental Protection Agency has standards for the amount and type of contaminants allowed at acceptable levels. One of the goals for these design solutions is to clean the air to at least the EPA standard, or even cleaner with the use of plant materials.

Excerpts from research by Federico Karagulian, et al., 2015, titled "Contributions to cities ambient particulate matter (PM): a systematic review of local source contributions at global level" helps to illuminate on air quality in urban areas. The research covers the time period between 1990-2014.

There are five categories of main sources of ambient particulate matter (PM) commonly found... and have been used for the purpose of this

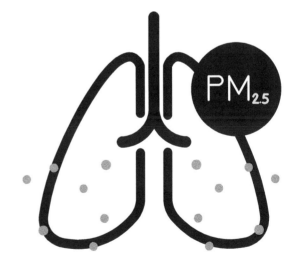

Image 1.12   The particulate matter (PM) in air pollutants is small enough to get into the lungs of anyone who is breathing the polluted air.

analysis: traffic, industry, domestic fuel burning, natural sources including soil dust (re-suspended) and sea salt, and unspecified sources of pollution of human origin.

- **_Traffic_**...includes different kinds of emissions from various vehicle types. In addition to primary PM emissions from exhaust, and the emissions of organic and inorganic gaseous PM precursors from the combustion of fuels and lubricants, vehicles emit significant amounts of particles through the wear of brake linings, clutch, and tires (Amato et al., 2009; Belis et al., 2013). These are deposited onto the road and then re-suspended by vehicle traffic together with crustal/mineral dust particles and road wear material.
- **_Industry_** is a category including mainly emissions from oil combustion, coal burning in power plants, emissions from different types of industries (petrochemical, metallurgic, ceramic, pharmaceutical, IT hardware, etc.) and from harbor-related activities. Industrial sources are sometimes mixed with unidentified combustion sources or traffic (Belis et al., 2013).
- **_Domestic fuel burning_** includes wood, coal, and gas fuel for cooking or heating such as in residential complexes.
- **_Natural sources including soil dust and sea salt dust_**. These components of PM are associated with the re-suspension from fields or bare soils by local winds. When reported separately, road dust was included in the traffic source category in this study (Belis et al., 2013).
- The "**_Unspecified sources of human origin_**" category mainly includes secondary particles formed from unspecified pollution sources of human origin. Primary particle emissions include mechanically generated particles and primary carbonaceous particles. Primary particles also include carbonaceous fly-ash particles produced from high temperature combustion of fossil fuels in coal power plants.

Additional PM are formed from reactions of gaseous pollutants, VOCs (Volatile Organic Compound), NMVOCs (Non-methane VOC), and evaporation of solvents such as paints, degreasers, and stain removers.

"Airborne particles like dust, soot and smoke that are less than 2.5 micrometers in diameter are small enough to lodge themselves deep in the lungs. Studies have linked pollution of this sort to respiratory problems, decreased lung function, nonfatal heart attacks and aggravated asthma, according to the United States Environmental Protection Agency." (The New York Times Article, A Study Links Trucks' Exhaust to Bronx Schoolchildren's Asthma, By Manny Fernandez, Oct. 29, 2006).

According to the New York State Department of Health and Department of Environmental Conservation "Particles in the PM 2.5 size range are able to travel deeply into the respiratory tract, reaching the lungs. Exposure to fine particles can cause short-term health effects such as eye, nose, throat and lung irritation, coughing, sneezing, runny nose and shortness of breath. Exposure to fine particles can also affect lung function and worsen medical conditions such as asthma and heart disease. Additionally, E.P.A. officials said these fine particles, a significant portion of which are produced by diesel engine emissions, lead to 15,000 premature deaths a year nationwide."

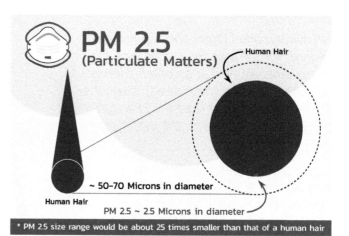

Image 1.13   Particulate matter 2.5, or more commonly known as PM 2.5, is imperceptible and small at approximately 2.5 microns in size. In reference to the diameter of the width of the hair, the small red dot is the size of 2.5 microns.

To understand the scale of 2.5 micrometers, there are 25,000 microns in an inch. The width of PM 2.5 is approximately 30 times smaller than that of a human hair. PM 2.5 is quite small, such that several thousand of them fit on the period at the end of this sentence.

In addition to necessary policy changes addressing gaseous and chemical release into the air, this chapter explores using plants to remediate for particulate matter (PM) in air. Some plants are able to take in the contaminants through their roots, stems or leaves and continue to live, removing the PM from the air so animals, children and adults can breathe in cleaner air with less PM.

SOUND WAVES

(Famulari, 2020)

The next item to address is sound. Urban sound is measured in dB(A), which stands for decibels with an A-weighting. This A-weighting includes an adjustment which takes into account varying sensitivity of the human ear to different frequencies of sound. A-weighted measurements consider the loudness, annoyance factor, and stress inducing-capability of noises with low frequency and moderate or high volumes. This also takes into account the very low and the very high frequencies and volumes of noise.

Why is this worth understanding? Understanding that sounds near airports, expressways, highways, freeways, trains, and subways have a higher dB(A) allows one to consider the next step of how to use design to lower the dB(A) for those who live or work in these areas.

Image 1.14 Sound coming from a source is directed toward a wall. When the sound waves reach the hard surface of the wall without interference from surfaces in their way, the sound waves are reflected back into the space. The effect can be loud and echoing.

Published research by Maarten Hornikx illustrates that plants can reduce the dB(A) ranging from 4 dB(A) to 7.5 dB(A). More information on this research can be found in the book credits. Dense foliage absorbs frequencies, while hard surfaces scatter them. Consider a sound frequency hitting a hard wall, or tree trunk- the sound waves scatter. However, with the softer dense surface of foliage, the frequencies are absorbed. The articles states that:

"Vegetation in urban areas has a range of ecological advantages. Vegetation acts as noise reducing possibility in inner city environments, in particular building envelope greening measures. Also, there is growing evidence that visibility of vegetation by itself affects noise perception positively. For inner city environments, vegetation types that can be considered are low-height noise barriers, vegetated façades, vegetated roofs and trees. Low-height vegetated noise barriers were shown to be useful in road traffic noise applications at street level. These devices can be placed close to the driving lanes, thereby yielding road traffic noise reduction of about 5 dB(A)."

Some municipalities, such as New York City, require trees as part of their development in order to assist with noise, views and air quality. Planting trees in New York City sidewalks has been the project of artists such as Alan Sonfist and Joseph Beuys. Even today, walking through New York City, these living artworks can be unrecognized by the community, yet, without these green artworks, the urban area would have deteriorated air quality and lack shade. Other benefits of street trees for a community include experiencing seasonal change, growth, and time changes through plant materials, as well as bringing people together to care for the living plant materials.

In looking at greenwall surfaces, it was found that the greenwall that does not have plant materials on it only reduces 1 dB(A), while a green wall with plant materials reduces 5 dB(A). It is the soft absorption of the foliage which absorbs sound.

Green roofs can decrease sound waves over buildings. It is the combination of plants, the sublayers of soil and liners, as well as roof shape which produce the benefits of sound reduction. It has been found that buildings with vegetated roofs can have an effect up to 7.5 dB(A) noise reduction.

"The effect of green-wall systems is larger for roadside courtyards than for trafficked streets and may amount up to 4 dB(A). An interesting application of vegetated surfaces are openings to courtyards, greening those surfaces has shown to amount to 4–5 dB(A) reduced sound pressure levels in the courtyard."

What does 5 dB(A) really feel sound like in volume and effects? Is it really significant? Yes! An increase of 10 dB(A) feels twice as loud to the human ear. For example, with a dishwasher with 50 dB(A) versus a dishwasher with 60 dB(A), the second dishwasher sounds twice as loud at the first one. 10 dB(A) feels twice as loud. Therefore, a decrease in dB(A) using plants and green design by 8 dB(A) feels almost half as loud.

The result- the more green, the lower the dB(A), which reduces hearing loss and stress, while creating healthy spaces in an urban area only creates positive outcomes.

Image 1.15   Sound waves go in all directions from the source of the sound. When it reaches a hard surface such as a wall, the sound reflects back into the space. Soft surfaces, surfaces with air holes or gaps, groups of people with space in between, absorb parts of the sound wave and reduce it before it gets to the hard wall surface.

PLANTS CAN ABSORB SOUND WAVES

(Famulari, 2020)

Plants can absorb sound through their leaves and foliage. Dense trees are more effective than trees with large spaces without greenery. Barks of trees are hard surfaces which sound waves bounce off and scatter. The green areas of plants are soft and have pockets of air. [Leafier plants are more beneficial for reducing soundwaves.] When sound waves are directed to the foliage, the sound waves are softened and absorbed into the leaves and air pockets, similar to a soft sponge. This is effective for urban areas where there are a myriad of sounds which move through the air. When dense areas of plant materials are in front of build-ings, properties, courtyards, yards, entrances or windows, the plants act as sponges for sound waves. These plantings reduce the sound that travels to residents before getting into the build-ing, property, courtyard, yard, or room.

Image 1.16 Interstate which has trees, shrubs, plants, and grasses designed along the sides and center. The greening reduces the sound waves from the vehicles to neighboring communities.

DISTANCE OF PLANTS TO SOURCE OF THE SOUND

Similar to the information with air contaminants, there are several factors in determining the distance of the source of the sound to see how far it can spread. The same factors include the current wind, the wind pattern, the structures which may be blocking the air passage (such as walls or buildings), the slope of the land, the current weather, and the frequency and sound wave caused from the source itself. Some additional [factors] which affect sound include the materials of the structures and frequency of the sound, as well as its volume.

Surfaces which are soft, dense and have air pockets absorb sound. Foliage and sponges can be used as sound proofing material in buildings. The frequency of the sound itself also affects how far it is carried. Low frequency sounds such as bass sounds can shake and vibrate much further than higher frequency sounds such as birds chirping. One other factor on how much the sound affects you is the context with which you hear the noise. For example, trucks moving in the middle of a day with other rustling traffic is less likely to bother you than a single loud truck of the same frequency in the middle of a quiet night. Though the frequency of the sound is the same, the context shift has a greater effect on you.

PSYCHOLOGICAL EFFECTS OF GREEN DESIGN

(Famulari, 2020)

Another item to explore in design

is the psychological benefits of views, green, and other designs which connect people to green nature. The [advantages] of green spaces not only provide [ecological] benefits on negating urban heat, offsetting greenhouse gas emission, and offsetting storm water, there are also health benefits of plant materials. From the University of Washington's Green Cities: Good Health, further mental benefits of green design include:

Image 1.17   The view of the Japanese garden from the interior spaces has similar psychological benefits to the being in the exterior garden itself.

- *The experience of nature helps to restore the mind from the emotional fatigue of work or studies, contributing to improved work operations and satisfaction.*
- *Urban nature, when provided as parks and walkways and incorporated into building design, provides calming and inspiring spaces and encourages learning, inquisitiveness, and alertness.*
- *Green spaces provide necessary places and opportunities for physical activity. Exercise improves cognitive function, learning, and memory.*
- *Outdoor activities can help alleviate symptoms of Alzheimer's, dementia, stress, and depression, and improve cognitive function in those recently diagnosed with breast cancer.*
- *Contact with nature helps children to develop cognitive, emotional, and behavioral connections to their nearby social and biophysical environments. Nature experiences are important for encouraging imagination and creativity, cognitive and intellectual development, and social relationships.*
- *Symptoms of ADD in children can be reduced through activity in green settings, thus "green time" can act as an effective supplement to traditional medicinal and behavioral treatments.*

Further information on the studies stated here can be found in the book credits. From the World Health Organization's, Urban Green Spaces and Health- a review of the evidence, their findings summarize that the:

…"available evidence of beneficial effects of urban green spaces, such as improved mental health, reduced cardiovascular morbidity and mortality, obesity and risk of type 2 diabetes, and improved pregnancy outcomes. Mechanisms leading to these health benefits include psychological relaxation and stress alleviation, increased physical activity, reduced exposure to air pollutants, noise and excess heat."

DEI IN DESIGN

One final important piece to consider is DEI in design. Green design can act as a form of activism and engaging the community. To be able to move freely, equitably, healthy, and safely, with open discourse and engaging the community, these are ideologies of design – all design – not just green design, landscape architecture, industrial design, etc. Design can act to inspire spaces for equality and engaging the community. This, however, is not abundantly the current practice of design in subtle and direct ways. Actively seeking to be inclusive in sustainable design is a newer approach in green design.

DEI is an acronym for diversity, equality and inclusion. Justice is also an important component of DEI. DEI is inclusive of a range of various people that are not typically the population designers traditionally construct for. In order for green design to be effective, reaching out to people in the community, asking questions, and doing research are vital for a design which is more encompassing.

Image 1.18 A pathway for people with different walking abilities is blended into the rhododendron public park in Helsinki, Finland.

Though landscape architecture has grown and evolved, the roots of it are still based in a non-inclusive past and the signs of that past are still visible in the present. Presently, public design in the United States is typically for 5'9" average height of men, visually able and able to freely walk, understand English at a high school level or higher, Caucasian male as the standard. This can be seen in the heights of signs, the types of signs placed, the language on the signs, heights of benches, and other standard site features.

In present-day New York City, parks in formerly red-lined areas, lower income areas, areas of housing projects, areas of primarily BIPOC residents, and areas of primarily immigrants with English as their second language are fewer and far between than those wealthy areas and areas of primarily Caucasian residents (Biggs, 2019; Gjording, 2022; Nardone 2021). Adding parks and green areas and retro-fitting these into lower income areas, formerly red-lined areas, and areas where the community is predominantly BIPOC or immigrants are important to address the disparity of shared greening.

Image 1.19 Green design under a freeway with urban agriculture.

Inclusivity is also about addressing minorities, womxn, people who identify as BIPOC, people with different physical abilities, people with different visual abilities, people with different walking abilities, people of varying heights not within the range of "average" adult (which is 5'9" for men and 5'4" for women in the United States), people whose first or primary language is not English, people who identify as LGBTQIA+, people who identify as non-binary, ease of access and use for single parents with multiple children, people with a variety of income levels, and a variety of other users. Inclusivity is a specialty of green design at best, and cursory or ignored at worst.

Equality is about design where it is available and designed for multiple users in equal experience and use, with discourse and shared experiences between mixed users. This ideology of DEI shows in site design through signage, seating heights, languages shown, cultural references, colors, scents, and narratives shown through the landscapes.

Some of the designs in this book are more successful in being inclusive than others. The sites in Chapter 2, for example, are designed specifically with DEI. In order to have healthy living designs across a majority of neighborhoods, using rooftops and vertical greening is an option for areas where street-level space is not available to retrofit greening. Plants which reflect the cultural

Image 1.20 Design for courthouse steps to create a pedestrian scale to an expansive structure.

diversity of neighborhood as well as a larger connection to other countries and cultures is also an approach to design.

In Chapter 3, the mixed use streetscapes explore improving DEI with adding planted areas to the majority of sidewalks. The width of the sidewalk is a primary factor in the success of the design. Where the sidewalks are wide enough, or a lane of the street can be rededicated to green design, these types of street greening may be successful. Planters at heights for a variety of users, such as children as well as people who use wheelchairs, is another application of green design. Empty spaces under freeways and buildings near freeways are familiar in areas with urban scars. Repurposing these sites with healthy shared space for the communities creates an asset for those neighbors that are separated by freeways.

The courthouse greening in Chapter 5 addresses DEI specifically because in American culture it has been discussed that justice is not balanced nor blind. The courthouse design changes the overpowering and intimidating scale of the courthouse through green design, to one which is pedestrian scale or the scale of an individual. In the examples of interior greening, DEI is explored by looking for design options for those people that do not have ease of access to be outdoors on a routine basis due to physical abilities or emotional states.

There is a long way to go in addressing DEI in green design and public design. It is a right for people to have healthy spaces to share time, dialogue, experiences, and events. Making sure spaces are truly inclusive is worth the continuing work.

MOVING FORWARD

Are these green designs really meant to be created as shown in the examples throughout this book? Yes! This book is not solely for opening your mind to ideas for your space or creating new policies toward healthier spaces. It is also to inspire you to gather a team of green designers and creative people to make it happen. The designs shown throughout this book are able to be created and based on existing designs or parts of existing spaces. The greening images are all shown to create a green healthy way of living.

Your site conditions likely have pieces which can be found in one or more of the green images in this book. Putting the pieces together so that green design can thrive in your sight is achievable with knowledge, design professionals, and creativity. The results of an integrated green design

Image 1.21   Green design for an urban neighborhood.

throughout smaller and larger sites, interior and exterior sites, densely built areas, and newly built spaces create a healthy living fusion for a thriving community.

REFERENCES

Amato, F., Pandolfi, M., Viana, M., Querol, X., Alastuey, A., & Moreno, T. (2009). Spatial and chemical patterns of PM10 in road dust deposited in urban environment. *Institute of Earth Sciences* Jaume Almera, Spanish Research Council (CSIC), C/ Luis Solé Sabarís s/n, 08028 Barcelona, Spain.

Belis, C. A., Peroni, E., & Thunis, P. (2013). Source apportionment of air pollution in the Danube region. http://iet.jrc.ec.europa.eu.

Biggs, C. (2019, May 10). Rooms with a view (and how much you'll pay for them). *The New York Times*. (retrieved 11 March 2023).

Cummings, B. E., & Waring, M. S. (2020). Potted plants do not improve indoor air quality: A review and analysis of reported VOC removal efficiencies. *Journal of Exposure Science and Environmental Epidemiology, 30*(2), 253–261. https://doi.org/10.1038/s41370-019-0175-9.

Famulari, S. (2020). *Green up! Sustainable design solutions for healthier work and living environments.* New York: Productivity Press. https://doi.org/10.4324/978042297434.

Fernandez, M. (2006, October 29). A study links trucks' exhaust to Bronx schoolchildren's asthma. *The New York Times*, Region Section.

Gjording, L. R. (2022, October 25). Redlining and its impact on New York City. *City Signal*, Real Estate Section. (retrieved 11 March 2023).

Karagulian, F., Belis, C. A., Dora, C. F., Pruss-Ustan, A. M., Bonjour, S., Adair-Roahn, H., & Amann, M. (2015). Contributions to cities' ambient particulate matter (PM): A systematic review of local source contributions at global level. *Atmospheric Environment, 120*, 475–483.

Nardone, A., Rudolph, K. E., Rachel Morello-Frosch, R., & Casey, J. A. (2021). Redlines and greenspace: The relationship between historical redlining and 2010 greenspace across the United States. *Environmental Health Perspectives, 129*(1), 1. https://doi.org/10.1289/EHP7495

Wolverton, B. C. et al. (1989). *A study of interior landscape plants for indoor air pollution abatement: An interim report.* Mississippi: NASA.

Chapter 2

Urban Areas

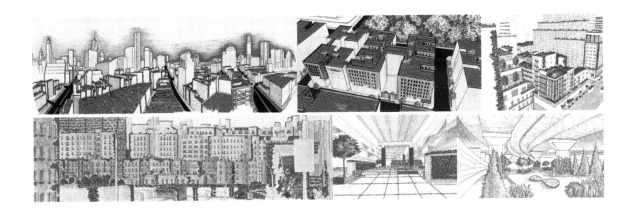

Image 2.1   Green design of an variety of neighborhoods and spaces.

Imagine if all city dwellers have views to an urban green space from their windows. This can be a park, courtyard, alley design, vertical greenwall, vertical garden, awnings with green design, street trees, sidewalk gardens, or a countless number of creative options. Greening design improves air quality, adds stunning views, improves the value of the real estate with green views, helps with emotional balance, and can be accomplished in new site designs as well as through updating existing spaces.

The designs throughout this and the following chapters explore these same basic ideas:

1 updating existing spaces for green design,
2 designing spaces with a goal of 10% of the horizontal and vertical surfaces for greening in order to improve the air quality,
3 creating designs in which city dwellers have an experience or view to green design from their window, for improving people's emotional health.

One focus of the green designs is to explore the percentage of greening to see if we can reach 10% of all surfaces to have planted designs. As greening designs cannot dominate the streets where there

DOI: 10.4324/9781003348634-2

Image 2.2   Panoramic view of downtown New York City taken from a 14th-story rooftop garden. The panoramic view spans to show downtown from the west side to the east.

is vehicular activity, the rooftops, terraces, vertical surfaces of buildings, and other urban structures become the best choice as locations for retrofitting greening designs. Rainwater as well as grey water from buildings is used for irrigation of the green designs.

Greening in an urban area adds to a higher economic status of a neighborhood. Homes with views to urban parks are more costly than similar homes without a green view (Briggs 2019). To help improve DEI, creating green access and views for all people across different social, racial, and economic groups, green design can help to positively improve communities.

The green design for this panoramic view of a downtown area focuses on horizontal areas primarily above the street level for the majority of green space and explores the large quantity of above-ground spaces which can potentially be utilized for greening design. Most buildings shown have additional vertical greening added in the corners of the building where there are no windows. With the goal of 10% of the horizontal and vertical space to have a living green design, the need for exploring the rooftops as the primary location for green design is an ideal option for urban areas.

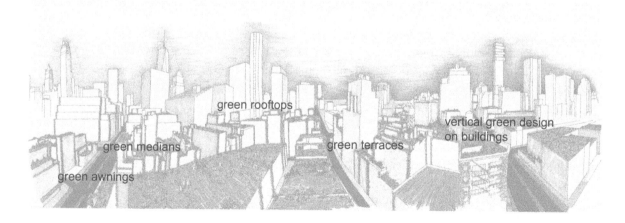

Image 2.3   Labeled illustration of the opportunities for green design in an urban panorama.

Image 2.4 Urban panoramic view with: a green design with select horizontal surfaces at street level and above street level with a living design; greening on building surfaces, corners, and spaces between windows; the large scope of living design allows residents and workers to experience or have views of greenings. Grey water from buildings as well as rainwater is used for irrigation.

With the density and varying heights of buildings, the use of vertical surfaces is required in order to get close to the 10% greening goal. The buildings which are above six stories are quite complex to retrofit the exterior of the building with green design techniques and can better incorporate green design within the interior of the building. Interior green designs are further explored in Chapters 3 through 5.

The density of the built space in urban areas, such as in Image 2.4 requires looking at both individual buildings and the totality of the buildings and structures in a neighborhood and community. The design in Image 2.4 explores the totality of the area for a design which reconsiders how a metropolis can be dense and green at the same time. This greening design serves a few goals: to improve air quality, and for residents and workers to have access to greening experiences either through being in a green space or having views to it from their windows.

The design of Image 2.4 is shown with a view primarily for the rooftops and nearby parks. This site can also have vertical greening design in the courtyards and exterior building surfaces. If most buildings have green designs on their rooftops and other surfaces, then the people in taller buildings have views to green design below their view. In lower buildings, parks, green alleys, and countless street trees allow residents to have views to healthy green designs from their windows.

The rooftop designs in Image 2.4 may or may not be accessible for gatherings of large quantities of people to spend time on depending on the structure of the roof. However, access for maintenance crews to caretake the roof is available. Grey water from the building and rainwater are the primary sources of irrigation.

Plants which are used in urban rooftops must tolerate higher winds due to their location above the ground without walls blocking wind. Plants which live and grow in direct sun are also required for this site design. There are plants and trees which are suitable for these harsh rooftop conditions as well as proper weather for the plant zone. The totality of the rooftop and green design reduces noise pollution as well as heating and cooling fluctuations within a building, which in turn reduces costs of the building.

The goal of 10% greening is not achieved in the design of Image 2.4 due to the quantity of vertical building surfaces. Air quality is still improved with this design as well as the heat island effect is reduced and the emotional health of people is improved. Exploring green design is through creating experiences and views for healthy living designs across an expansive area and community rather than selectively in wealthy neighborhoods.

As the first design of the downtown panoramic takes on a more expansive scale, the second design, Image 2.13, design explores one block in the South Bronx. Two six-story buildings take up most of the square block in this neighborhood and are the focus of green design. The majority of the ground is covered with roads and sidewalks; therefore the rooftops in urban areas can be used to improve the air quality as well as enhance views from people's windows.

There are numerous studies (Nuccitelli, 2023) which state that views to parks, gardens, forests, and green spaces improve people's physical and psychological health. These green designs explored within this book address what these green expansions can look like on different scales and locations.

Expanding on the idea of city residents and workers having views to greening, it would require that a substantial amount of buildings have green designs on their rooftops. This would provide users of taller buildings the opportunity to observe green design on all available surfaces

Image 2.5   The Bosco Verticale, located in Milan, Italy, designed by Stefano Boeri.

Image 2.6   Translated to "Vertical Forest," the Bosco Verticale is a vertical urban forest.

Image 2.7   The Bosco Verticale residential building in the urban area of Milan has interior and exterior greening through the design.

Image 2.8   Throughout the year the plants in the Bosco Verticale lose foliage, grow foliage, and change colors.

Image 2.9   Google Earth image of six-story buildings in the Bronx, New York.

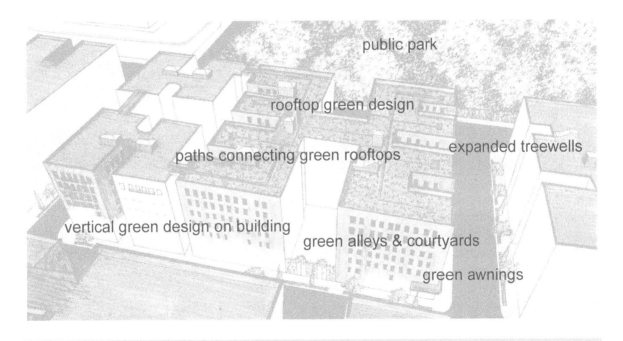

Image 2.10   Labeled green design of a building in the Bronx, New York.

below. For people who reside or work in buildings with fewer floors, views to windowsill planters, planted awnings, planted gates, green alleys, planted court-yards, fire escapes, patios, street trees, sidewalk gardens, street planters, greenwalls on buildings, and public parks can provide sources of green-ing for urban residents and workers.

The block scale design (Image 2.13) shows green-ing on a variety of layers and scales. A park, roof-tops, street trees, sidewalk gardens, green courtyards, green alleys, planted gates, and windowsill planters are all designed to fit into the existing block. This allows all residents to have views to green design from their urban studios. This block of green design, though falling short of the 10% goal, still improves air qual-ity and emotional health. The addition of vertical greening on structures and buildings will be needed to reach the 10% greening goal. An effective biophilic design for this site incorpo-rates horizontal and vertical installations of greening techniques from the street level through to windowsill planters to the rooftop.

Image 2.11   Green roofs on urban buildings.

Image 2.12   Green roofs on modern buildings in Sydney, Australia.

This site example focuses on the large quantity of surfaces above the street level which are avail-able for green design. Street-level surfaces do not accommodate large-scale greening on this block; therefore, street trees and sidewalk gardens are added on the ground plane.

Image 2.13   A greening design for an urban block with: green design on rooftops; pathways across the buildings connect the structures to create a larger green scope; nearby park has hills with trees which grow as tall as the nearby buildings, creating a visual connection of a wave of living green design; vertical building surfaces, awnings, alleys, courtyards, and sidewalk gardens and trees; residents, visitors, and workers experience and view greening throughout the neighborhood. Illustrated for the rooftops and nearby park, this site can also easily have vertical greening design in the courtyards and exterior building surfaces.

Not all rooftops are structurally suitable for large quantities of people to spend time on due to the load-bearing weight of the roof. Green design can be created with modern lightweight materials such as lightweight nutrient-rich engineered soils which are available and improve the range of designs for rooftop greening. While it may not be suitable for large quantities of people, maintenance crews are provided access to the roof and green design which uses rainwater and the buildings' greywater to irrigate.

The lack of ability to walk on rooftops with green design does not diminish the value of the green design to the people, or building structure and larger scale. On an individual level, emotional and physical health is improved with views to green while also improving air quality (Biggs, 2019; Nuccitelli, 2023; WHO, 2016; Wolf, 2010). On a building level, the addition of plants on a roof, whether a sloped roof or flat, keeps the roof temperature from fluctuating too severely in the extreme weather conditions of a hot summer and cold winter. The planters, soil, and plants keep the roof and building more insulated and keep the temperatures from extremes, thereby lowering the cost of heating and cooling over time.

On a larger scale of the block, a neighborhood and city can improve air quality and living spaces. The addition of living green designs creates changing views with plants growing throughout a year and decades. Plants also improve air quality by adding oxygen to air for people to breathe. The green block design (Image 2.13) improves the urban fabric for the people, animals, and air health and use rain water and grey water.

Image 2.14   ACROS Fukuoka Prefectural International Hall in Tenjin Central Park, Fukuoka, Japan with plants growing on multiple stories of the building.

Image 2.15   Rooftop garden in urban setting with trees, areas of shade and seating.

Utilizing both horizontal and vertical surfaces provides green design opportunities both at street level and above. Image 2.20 showcases how green design can enhance the experience for pedestrian and building users alike. There are a myriad of vertical surfaces that are utilized for green design in this neighborhood in NYC.

In this area of New York City there are buildings of varying stories – from five story walk-ups to countless stories; varying in scale from the width of 1 studio across, to buildings with multiple rooms on a floor; and varying in age from the early 1900s to modern architecture.

Image 2.16   A street corner in downtown New York City with buildings of varying stories, scales, and ages.

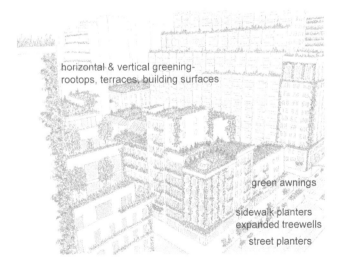

Image 2.17   Labeled design of a green corner in New York City.

Image 2.20 illustrates how on the street level people can enjoy expanded tree wells which are more than tree protection measures. Expanded tree wells can provide more expansive spaces to accommodate layers of smaller plants, trees, and seating. This design practice is already utilized throughout cities as can be seen in Images 3.12 and 3.13. In addition to expanded tree wells, median street planters are also created for the street-level experience. Street planters are designed to divide one lane of the road to allow cyclists to ride with a safety divider of a flow of green along their ride.

Image 2.18   An area of the street protected for pedestrians by large planters.

Slightly above the ground plane while still on pedestrian scale and visible are awnings above first floor retail which are planted with small green gardens. Above the street there are windowsill planters on five story walk-up buildings, and garden terraces on taller buildings. Vertical greening on buildings is celebrated with colorful plants appropriate for urban conditions. Vertical greening can also improve air quality while creating a unique and vibrant urban jungle. Throughout the neighborhood green design can be observed and built on the rooftops of buildings of all heights.

Image 2.19   Living green design on building front.

Rainwater and grey water from the buildings are both recycled for irrigation on site. Pipes can be added on building surfaces for both the plants to grow as well as for irrigation. The pipes direct the rainwater and greywater to the areas which need it. Plant selection for the urban design is not only specific to the plant zone but also ones which can thrive in direct sun and live during times of low water and higher rainfall.

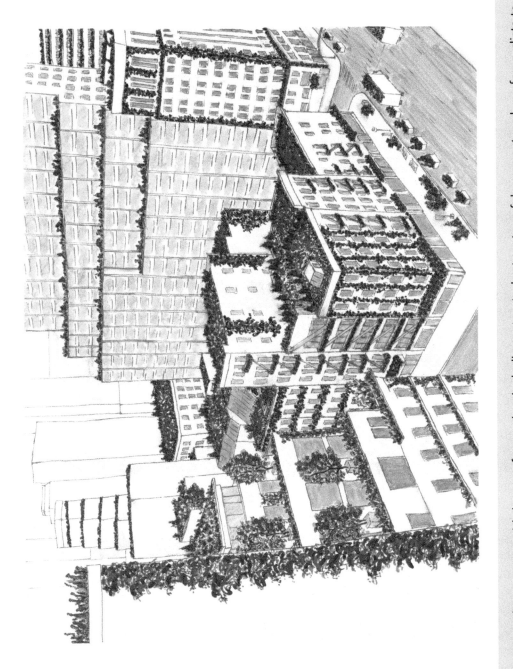

Image 2.20   An urban corner with layers of greening including: street planters to safely separate a lane for cyclists to ride with a green edge; sidewalk planters include trees and extended areas with plants and benches; horizontal and vertical greening of building surfaces include of awnings, corners, between windows, terraces, and rooftops; grey water from the buildings and rainwater is used for irrigation of the site; residents, visitors, and workers can experience and view greening while the urban air quality is improved.

This urban corner design (Image 2.20) begins to get closer to the goal of 10% minimum of green design. However, due to the height of the taller buildings, this site does not meet this goal on the exterior of the site. Though, with the addition of interior green designs (seen more in Chapters 3 through 5), the overall totality of the greening is a great progress in urban health. The green plan in Image 2.20 continues to explore design as a form of improved health, both emotional and physical, for the people to reside or work in the area.

Image 2.21   Green roof and vertical greening in urban area.

The integration of urban greening at this scale and depth addresses an approach that people in onlooking studios, lofts, commercial spaces, or pedestrians at the street level can see or experience green design from where they stand. Urban design provides a great opportunity for unique, progressive, and advanced green design. The opportunity to create a design on multiple scales, levels, and locations can be achieved by looking at site conditions and spaces for greening.

Image 2.22   Ecology and green living together in an urban setting.

Health is a right, not a privilege, and designing with this expanded approach addresses one part of this right. Addressing DEI in design includes creating green spaces for people in communities of mixed economic levels, mixed abilities, mixed culture, mixed genders, and people of mixed races. Designs in this chapter which span a community explore DEI.

Image 2.23   Photograph taken from the street at eye level. The South Bronx is a unique neighborhood and is home to an art museum, residential buildings, storefronts, mixed-use buildings, parking garage, and an above-ground subway.

Shown in Autumn, this South Bronx streetscape Image 2.27 uses small, detailed greening interventions as well as larger surfaces for a sprawling design. The totality of the green design benefits users of the space with access to views in a space retrofitted with a healthy mix of dense structure and live plants throughout the community.

This design includes street trees, expanded sidewalk gardens with seating and plants, green terraces, greenwalls on buildings, green awnings, green roofs on the above-ground subway, green roofs throughout the neighborhood, greening which grows over the roof, balconies, terraces, and windowsill planters.

In this example, the totality of the streetscape green design accomplished both goals of meeting the 10% greening minimum while also providing onlookers with a view of green design. With the totality of the streetscape, the goal of having residents, visitors, and workers have a view to green design as well as having 10% of the space green is achieved.

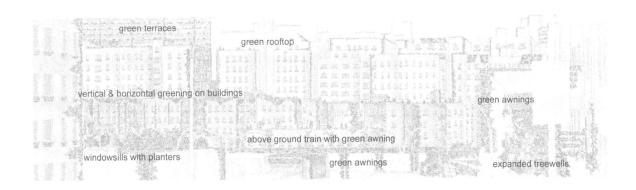

Image 2.24   Labeled design for a street scene in the South Bronx, New York.

Starting at the street level, tree wells with expanded gardens are utilized throughout the neighborhood. Window planters in Image 2.27 are designed for the buildings for both interior and exterior plants, allowing plants to grow more consistently throughout the year. Moving from the street level, vertical greening on the building edges and between windows or on structures protruding from the building face provides ample space to retrofit green design in the existing infrastructure. The styles of these structures can vary from mechanical pipes to ornate and stylized structures. The conclusion is this neighborhood is suitable for these green interventions.

Image 2.25   Urban green design in London, England.

Creating the design with images from the ground plane or pedestrian view is beneficial. This allows people to design spaces looking at it from the way we walk, live, see, and experience them, allowing the users' point of view to be clearly shown. In this in the design Image 2.27, the view from a street in the South Bronx is shown for how people envision this neighborhood can be and how creative individuals can visualize it in their minds on an Autumn day.

Image 2.26   Ecological building façade in Danang City in Vietnam.

Greening is designed through layers and scales of the neighborhood, including the street level as well as above the eye level (what you see when looking up), and on both horizontal and vertical surfaces. Building from the concept that people should be able to experience or see greening from their window or view, the integration of green design at a neighborhood scale becomes more inventive, particularly when retrofitting an existing space.

Image 2.27   An illustration from the pedestrian view (at eye level) allows people to see the site as they would experience it while walking. This greening of an urban neighborhood includes: expanded tree wells with trees, plants, and seating; vertical and horizontal building surfaces such as building corners, windowsills, awnings, terraces, and roofs; illustrated in Autumn with blazes of oranges and red flowing through the urban fabric; rainwater and greywater are used for irrigation of the neighborhood; residents, visitors, and employees experience improved air quality with green spaces to walk and see on multiple heights and planes.

Designs which are drawn from the perspective of the individual experience, such as in Image 2.27 are beneficial for understanding what the experience will be for the pedestrian. This helps people to put themselves in the spaces.

Image 2.28   Vertical garden on high-rise building.

Image 2.29   Vertical greening on high-rise building.

The areas under freeways, interstates, and highways are opportunities for unique green design interventions which can engage the people within the community. Some communities have utilized these spaces for shopping areas, green design, parks, and much more.

Two green designs in this chapter are spaces under raised roadways, Images 2.34 and 2.41. The greening design in Image 2.34 turns the spaces under the highway into an area for urban agriculture using storage containers to create horizontal and vertical spaces for edible plants. Seating, tables, urban agriculture, areas to clean the produce, and gathering spaces to have workshops are displayed in this design. This showcases an example of using underutilized spaces within our cities and bringing them to life, while also improving air quality in the urban area.

Existing architecture in a downtown space and forgotten or lost spaces within our cities can be reclaimed for the benefit of the community. There are numerous opportunities for used and typically unused urban spaces where green designs can be applied. Alleys, underpasses, areas under freeways, courtyards, under bridges, as well as vertical surfaces and spaces are areas for neighborhoods to explore greening for their sites. The incorporation of green design can provide additional benefits to these spaces and communities.

Image 2.30   Under highways, freeways, and interstates, there is a great opportunity for green design for the people of the community.

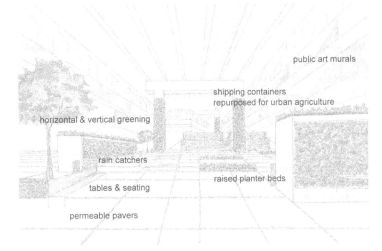

Image 2.31   Labeled design for urban agriculture under a highway.

Urban agriculture is integral in the design of Image 2.34 using shipping containers as structures and repurposing them from their original use. Shipping containers are currently being reused for housing structures with single and multiple units being placed together for site designs.

In this green design installation example of Image 2.34, agricultural use for the shipping containers, with an altered sliding opening of a clear front side for sunlight and visibility. This helps neighborhoods develop homegrown produce for community members and provide food support and growth. Community partners who care for the site can invite volunteers as well as provide paid opportunities to tend to the produce while learning practices.

The example in Image 2.34 has a variety of features: urban agriculture, permeable pavers, seating, tables, water features, artists-created rain barrels, use of rain and grey water for irrigation, artist-created murals on the shipping containers and sides of the freeway, greening vertically on the columns, greening on the top of the shipping containers, plants which reduce sounds from the surrounding areas, and gardens for children and people with different walking abilities.

Image 2.32   A lush green design under a metro bridge.

Image 2.33   Corrugated grey water storage tank runoff reservoir collection container.

Image 2.34   An urban agricultural design under a highway includes: permeable pavers across the site as well as raised planter beds and seating; shipping containers are repurposed for growing food for the community as well as plants on the roof; structures have vertical greening throughout the site; rain catchers located throughout the site are used for irrigation in addition to fresh water; public art throughout the site on rain catchers and freeway surfaces which celebrate the community.

Shipping containers come in three standard sizes: 8' width, by 8'6"–9'6" height, by a choice of 10', 20', or 30' length. Shipping container gardens and housing are currently part of site designs throughout the world. There are shipping containers which have been used to create buildings, homes, gardens, restaurants, medical clinics, and much more. Windows, doors, solar panels, and other pieces are added to create a comfortable space.

Permeable pavers come in a myriad of sizes and styles and are used through the site design in Image 2.34 to allow water to return to the earth rather than end up off site in a storm drain. Further information about permeable pavers can be found in Chapter 1.

Moving up from the ground, the design includes greenwalls and greening on the tops of the shipping containers utilizing a range of surfaces which can be designed with plant materials safely. The greenwalls and all plants are irrigated with rainwater and greywater.

Rainwater as well as grey water from surrounding buildings can be used for irrigation of the site. Recycling this water allows fresh water for other needed uses. Rain is collected in colorfully painted rain barrels throughout the site near columns and structures. There are numerous international sites which celebrate the arts publicly. Philadelphia, Pennsylvania, celebrates art throughout their city with murals which show the rich stories of the community. Bringing artists to this freeway green site to design rain barrels, murals, and storage container artwork celebrates not only the artists but the community they are working in.

Image 2.35   Shipping containers designed as housing with two or more containers placed together and stacked. Solar panels are added to create a sustainable form of energy.

The neighborhood buildings have terraces, fire escapes, greenwalls, and roofs which can accommodate green designs. As noted earlier in this chapter, rooftops with green designs have benefits for the building and the community of which they are a part. The totality of the integration of green design components brings this neighborhood closer to the 10% greening goal.

Image 2.36   Mural on South Street in Philadelphia, Pennsylvania, celebrates the community.

Freeways are common occurrences which have created under-utilized underpasses which can be designed for people and greening. The site in Image 2.41 illustrates an underpass at dusk that incorporates green design techniques. The application of green design in this example provides solutions to multiple common urban concerns. As shown, the green design reduces noise from overhead and nearby vehicles; it provides additional public green space with seating; utilizes both rain and grey water for irrigation purposes; and creates vertical growing surfaces on the support columns.

As an aesthetic, the reflective water feature, public murals, and public art of the constellation of lights can be enjoyed by the people of the community. The effects of having healthy green spaces for improving emotional and physical health are discussed in further detail in Chapter 1.

Image 2.37   This area under an urban freeway can be redesigned as a green space with public access for the community.

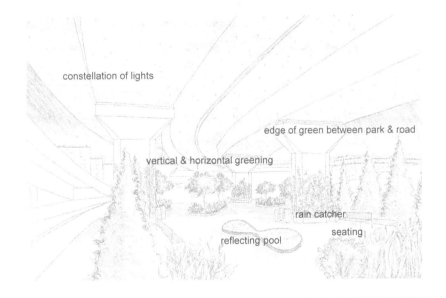

Image 2.38   Labeled design of a park under a freeway.

Suitably selected plants can provide an effective design solution to reduce sounds across a site. Understanding sound waves, how they work, and how plants can help to reduce sound waves is beneficial. More information about sound waves is in Chapter 1.

The public art in Image 2.41 includes LED lighting underneath the roadway which is inspired by the constellations. Due to light pollution in urban areas, large amounts of stars are not typically visible. This artwork uses LED lights which can be programmed to have constellations throughout the year. The public artwork becomes an educational tool for the people who visit the site. Plants are designed to be placed in a variety of sizes of planters on the ground. Rain chains are used to move water throughout the site with colorful rain barrels for collecting rainwater for irrigation of the plants.

The reflecting water pool with an infinity symbol shape is near the seating. The infinity design uses shallow recirculating water which creates a slight warping and moving

Image 2.39   Underneath an elevated highway in Shalu, Taichung, Taiwan, is a boardwalk through a park.

Image 2.40   A playing field under a freeway in an urban design.

effect in the reflection. Sitting in a space with a reflective piece, lighting of the constellations, and luscious green design encourages people to pause for a time and just be present. Plants with soothing fresh scents are designed into the site to encourage people to take a time to pause. The reflective feature and scented plants in the design help people to sit, slow down, and pause while looking at the reflection in the water; this can be a much-needed pause for people.

Image 2.41 A park under a freeway includes: LED lighting under the freeway inspired by a constellation of stars which can be programmed to change revealing different constellations throughout a year; trees and leafy plants reduce sound waves from the nearby traffic as well as improve the air quality; seating for groups of people as well as space for activities; rainwater and grey water from nearby sites for irrigation; a reflecting pool near the seating which reflects the constellation.

In this design, plants and hardscape are used to screen the amount of sound and light that can carry from one property to another. Buffers can also ensure attractive views of a property and to neighbors. Appropriate landscaping also moderates temperatures of impervious areas, reduces glare from parking lots, and helps filter automotive exhaust.

(Famulari, 2020)

Buffers and screens are used to reduce the impact of use of land on adjacent areas, which can be significantly different in character, density, or intensity. A buffer consists of a horizontal distance from a lot line. This buffer can consist of screening, landscaping, and fencing materials. Screens can be made from landscaping, fencing, berms, or combinations.

(Famulari, 2020)

Image 2.42   A ping-pong table in a park under a freeway.

Green buffers create a different space within the green design of the park under the freeway. People are encouraged to have activities such as gatherings, shows, workshops, running, strolling, and others in this green design.

This greening freeway park uses plants to create a comfortable pedestrian scale space while improving air quality, adding an urban greenway, and reaching the 10% greening goals. With the scented plants and tall grasses moving in the wind, reflections of the lights, seating, and open space, the community is able to have more spaces for events and public use.

Image 2.43   In Taipei, Taiwan, a highway is designed with a park system in the urban area.

Image 2.44   A vertical wall in São Paulo, Brazil, is planted with a living green design throughout an expanse of the roadway.

The designs in this chapter and throughout this book celebrate people living and working in healthy spaces, both interior and exterior sites, including greening in their residences, commercial spaces and public spaces. In the images shown in this chapter, people in neighborhoods experience green design while walking, sitting at home, or visiting friends and family. This chapter celebrates urban green design as joyous sharing of culture and community through our shared spaces, views, and surroundings.

Having healthy spaces to live and work is a basic need and can be improved with the 10% greening approach to new and existing site design. As people continue to approach sustainable design with innovation, new technologies, and shared dialogue between fields, the directions people can take toward healthy communities will continue to grow exponentially.

REFERENCES

Biggs, C. (2019, May 10). Rooms with a view (and how much you'll pay for them). *The New York Times*. (retrieved 11 March 2023).

Famulari, S. (2020). *Green up! Sustainable design solutions for healthier work and living environments*. New York: Productivity Press. https://doi.org/10.4324/978042297434.

Nuccitelli, D. (2023). *The little-known physical and mental health benefits of urban trees*. Yale Climate Connections. https://yaleclimateconnections.org/2023/02/the-little-known-physical-and-mental-health-benefits-of-urban-trees/. (retrieved 12 March 2023).

Wolf, K. L., & Flora, K. (2010). Mental health and function-A literature review. In *Green cities: Good health*. College of the Environment, University of Washington. www.greenhealth.washington.edu.

World Health Organization. (2016). *Urban green spaces and health*. Geneva: WHO Regional Office of Europe.

Chapter 3

Commercial Spaces

Image 3.1   Green designs of storefronts, street scenes, commercial spaces, parking structure interior and exterior, and a parking plaza.

Commercial spaces vary greatly based on the service of the spaces, the users of the spaces, the needs of the business, short- and long-term users, and site conditions. The differences between storefront, mixed use commercial street spaces, shops, plazas, and parking structures in this chapter illustrate a range of spaces with greening opportunities.

The spaces in this chapter are for multiple styles of communities: from smaller towns to urban areas, from multiple story parking garages to community parking plazas, and from retail shopping to smaller family-owned shops. The overall understanding is the same with a goal of adding green design to 10% of the surface areas to improve the health of the people who use the space, as well the community of people who live and work in it.

DOI: 10.4324/9781003348634-3

As a reminder, the designs throughout this and all the design explore three basic ideas:

1 updating existing spaces with green design,
2 designing spaces with a goal of 10% of the horizontal and vertical surfaces for greening in order to improve the air quality,
3 creating designs in which people have a view to green design from their window for improving their emotional health.

Image 3.2 A commercial space with a large window front can be an opportunity for adding an interior greening design to the space.

The designs in this chapter vary in scale and complexity, with some being able to be created by individual store owners and some spaces requiring contractors and permits to redesign larger sites. The spaces in this chapter include shops, and plazas. Shops are diverse in their use and business type. Flexibility with some stability for long-term growth is the approach to the green design of these spaces. The rewards are green designs which flourish allowing time to be seen through the gardens and green spaces as well as a sense of peace while caretaking for the space. Plants are living artworks and add life to spaces with the changing scale, colors, and shapes – let alone their ability to improve air quality, provide fresh scent, and calm one's soul just by being in the green space.

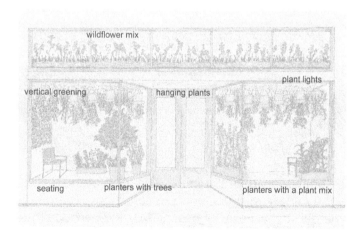

Image 3.3 Labeled design of green storefront window.

A street front store window is an opportunity to create living displays which are green all year round on the interior of the space which has a constant temperature inside. The storefront green design in Image 3.4 is inviting shoppers all year round with a living design that flows throughout all the windows of the space. There are a myriad of interesting details which are included in this space. These include upside down planters, flowering greenwall, plant lights added to the ceiling of the windows, grasses, vines, trees, and a layer of small plants with the soil and plant roots visible through the window.

Image 3.4 Storefront window with: A window front with hanging plants and herbs as well as wildflowers, grasses, and trees; irrigation for the hanging plants, floor planters, and wall planters from wires and lines attached to each planter; a vertical green design behind windowfront seating for multiple people; plant lights are added to the windowfront for successful plant growth; a shorter window with wildflowers, and the roots of the plants can be seen through the glass; greenwalls on the outside of the building can extend the greening to both interior and exterior of the site.

Commercial businesses with window front spaces have an opportunity to continually invite people into their business with a welcoming green window. With the live window display there is fresh air (Image 3.4) and changing scene of growing plants and blooms. This green entry area with different retail business inside attracts people to experience changing green design which grow throughout the year no matter the weather outside. In colder months, a warm area with green and flowering plants is an inviting site as well as a place for people to sit and stay inside the business for an extended time and talk.

The greening design is a benefit for the employees as well as visitors. Exposure to living green design

Image 3.5   Plants growing with roots visible.

Image 3.6   Plant growing in an upside-down planter.

improves memory as well as emotional well being (Raanaas, 2011; Ulrich, 1991). The addition of a space for a full green design where all employees can experience fresh air and scent, no matter the outside weather, improves people's health – both physical and emotional. With employees who are on site for hours each day, their physical and emotional health is strengthened with a green design in the window.

Image 3.4 has plants which vary in scale, color, textures, and scent throughout the entry design. A greenwall is added to the side wall, planters are added to the floor with structured plants and trees as well as loose plants that move with a breeze to create a variety of shapes and forms.

Plants are hung upside down from planters. There are commercial planters which are designed to grow plants upside-down with smaller plants. Inverted planters can also be made in a variety of sizes for plants to grow upside-down for a unique display. Looser hanging plants can be added with a similar effect.

Lights with a full spectrum white light or plant lights are beneficial for the display for healthy plant growth. Though this area has a large window, the cardinal direction of the window as well as awnings may not allow enough sunlight into the site. Plant lights or full spectrum lights on a timer can help to add needed light to the design.

Image 3.7   Plants grown in an upside-down planter.

Image 3.8   A building in an urban area with plants growing on the sidewalk and upward on the building.

In the smaller windows above the main area there are smaller colorful plants. A design in which the soil is visible with glass planters is one option for the plant's root to be visible and makes the processes of plant growth more apparent and celebrated. Covering the soil with opaque planters is also an option for the space.

Commercial spaces are also on streets with shops and private residents. This is called mixed use. On a block scale, urban spaces have a multitude of opportunities in their mixed use areas to experience unique greening designs in both interior and exterior sites. The street setting in Image 3.11 can be seen almost in any city or town. With mixed use buildings and a variety of people who live and work here. A universal approach of layers of small-scale green interventions and large-scale greening is applied throughout the horizontal and vertical spaces of the block. A public walkway can combine bus stops, walking spaces, bike spaces, seating, public greening, and other needs successfully. With the use of the exterior and interior green design the goal of 10% greening is achieved in this block.

Image 3.9   A street in downtown with mixed-use commercial spaces, offices, and residences.

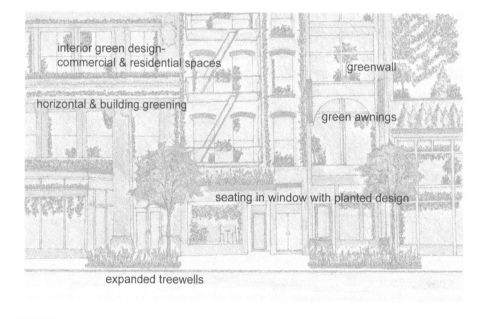

interior green design- commercial & residential spaces

greenwall

horizontal & building greening

green awnings

seating in window with planted design

expanded treewells

Image 3.10   Labeled design of green block with mixed use spaces.

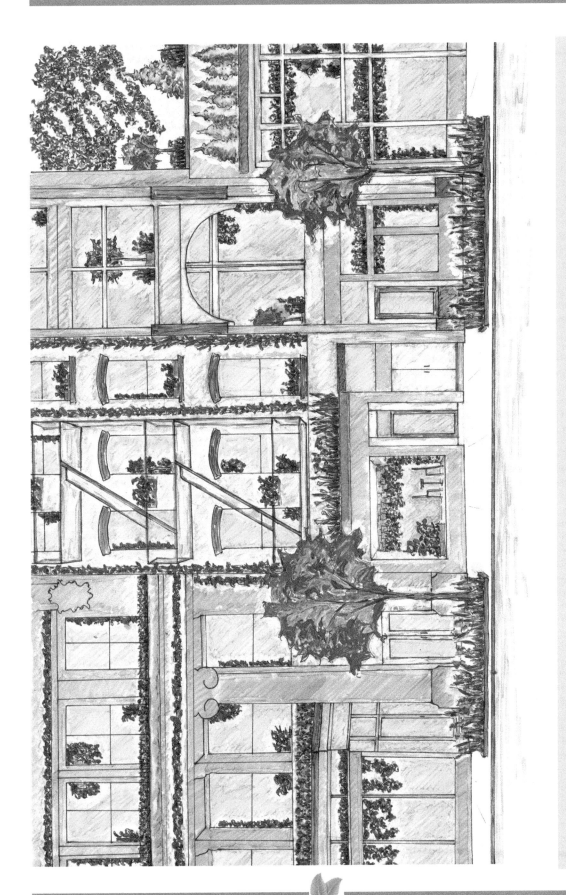

Image 3.11   Green design for mixed-use streetscape as seen from eye level or pedestrian view: interior and exterior greening for residences and businesses; street tree wells are expanded for fuller sidewalk living designs with plants and seating; greening for building surfaces and awnings; green designs for residence and business window displays; seating in windows for winter use and all year round; grey water from the building is used for irrigation of the green design.

In Image 3.11, tree wells are expanded to spaces with larger gardens and seating throughout the block. There are multiple benefits to the expanded tree wells. These designs expand the soil space which provides more room for plant roots to grow as well as alleviates compacted soil which is common in street tree pits.

Image 3.12   View of a street in New York City with expanded green tree wells.

These green sidewalk areas also allow for water from rainfalls to be utilized for greening rather than the water flowing to sewers.

Windowfronts of the spaces of Image 3.11 have plantings from the ceiling and side walls which allow the center of the window for other types of display or even for people to sit. This can be beneficial if the site is a café, bookstore, or other types of stores where an individual may want a bit of time to sit and enjoy the fresh air and green views within the store interior all year round.

Awnings, ornate building features, and ledges of the buildings have plants which can grow in shallow horizontal planters with soil. Greening is also designed on building fronts and building sides with greenwalls and patterns of greening along vertical building spaces. Window planters are designed for interior and exterior use with plants

Image 3.13   The expanded tree wells now are sidewalk gardens with seating and layers of plants.

on both sides of the windows. Additionally, interior spaces have gardens as unique as their users including green walls, trees, hanging gardens, children's gardens, and much more.

Grey water and rainwater can be used for the green design of this block in Image 3.11 with interior green designs also being utilized for irrigation. The greening of the streetscape improves the air quality and meets the goal of having 10% of the surface space to have living green design.

Image 3.14   Plants growing on top of the bus stop structure.

Utilizing a small- and large-scale green approach has multiple benefits such as retrofitting a street to be pedestrian scale with a greater comfort and ease of walking while having streets for vehicles. Additionally, some cities are reducing vehicle lanes through zoning practices and adding protected bike lanes and public transportation lanes in lieu of vehicles. The designs are created from the perspective from which people walk and see the space, thus encouraging city planners and business owners to directly see the effects of green design in their view.

The design in Image 3.11 and others throughout these chapters are spaces which have the ability to use green design in both newly built spaces and existing spaces with an updated design. The scale of the site and the options of green design are not limiting and instead are opportunities to be creative and see the spaces for how they can be with healthy green options.

Image 3.15   The mural in the urban area intentionally uses the vines growing above the wall as a living component of the artwork allowing it to drape over the portrait as hair.

The green design for the shopping plaza in Image 3.18 is shown in a winter magic after a light snow and the sidewalks have begun to be shoveled.

This shopping plaza has structured awnings above the windowfronts of each store. There is a great opportunity to add stunning live displays for long-term growth which are not only beneficial for improving the air quality of the shopping plaza with surrounding parking lots but also for an inviting display of year-round plaza greening.

Image 3.16 A shopping plaza with a similar repetitive design for the stores is an opportunity to have an extended green design above the storefronts.

These green designs create an attraction for business owners and customers, as well as people who work in the businesses or shop in the stores.

awning with living design

greenwall

interior greening

exterior planters on building

Image 3.17 Labeled green design for a shopping plaza.

Image 3.18 Plaza of stores with green design including: the awning above all stores has full living designs, illustrated during a light winter snow; exterior greening continues on building surfaces between stores; interior greening in stores with air plants, planters, and trees with artificial plant lights as needed; the designs create the feeling of spring on the interior and winter on the exterior; -the plants improve the air quality with fresh air both inside and outside.

The shopping plaza in Image 3.18 has structured awnings above all the storefronts. There is a great opportunity to add remarkable live displays for long-term growth which are not only beneficial for improving the air quality of the shopping plaza with surrounding parking lots but also for an inviting display of year-round plaza greening scenes using rainwater and snow. There are large greenwalls on the building front as well as between stores. Inside the stores the greening displays thrive throughout the year.

In areas of color weather, during the winter these awning designs celebrate the magic of year with fresh white snow collecting on the planted scene. Access and care for the overhead structure can be gained from the upper-floor store windows and doors or with a ladder from the sidewalk. In other times of the year, the magic of the fresh plant growth with new leaves arrives in the spring; blooming plants in the summer; and changing colors with deep reds and oranges in the Autumn. With the series of awnings across the plaza, the full scope of the space can be seen as a larger green design. With the added use of the rooftop for greening, the plaza becomes a space which can improve the air quality with building green design and an open green parking plaza.

Image 3.19   Urban street with green design on a structured awning

Image 3.20   Maintenance of the green design is gained by ladders from the sidewalk.

With the large awning design, the large greenwalls, the smaller greenwalls, the green rooftop, and the interior greening, the goal of 10% of the surface area to having living green design is successfully achieved and improves the air quality.

In order to create the layered green design in Image 3.18, the space on the building between the stores is utilized for exterior eye-level design. Layering a design means creating a design which looks at the ground plane, eye plane, and overhead plane, as well as designing the foreground, mid-ground, and background of the space. In this design, the plants on the ground are inside the stores, while the plants at eye level are the larger greenwalls and the green spaces between the stores and inside the stores. The over-head plane is designed through large green awnings and rooftop plantings.

Image 3.21 Plants growing from a structure which hangs from the ceiling.

The large greenwall and plantings between the store windows add vertical-ity to the green design uti-lizing more exterior surface area for greening the plaza and using plants to improve the air quality while making a larger green plaza design yet still at pedestrian scale.

Inside the stores there are interior green designs with trees, grasses, leafy plants, and hanging air plants. While outside the tempera-ture changes, inside the store the temperature is more consistent allowing for different interior plants to grow all year round. The contrast of the snow scene on the outside with the spring green design inside creates a rich scene for all employees and visitors to enjoy. The difference in the time of year will also be a dazzling design which worked and visitors will enjoy.

Image 3.22 Tree growing inside a space with a large amount of natural sunlight from the ceiling light.

The design in Image 3.25 uses pedestrian-scale greening on the street as well as vertical greening throughout the building. A multiple-story parking garage has walls with open surfaces for air circulation for vehicle exhaust to easily move through the space. A plant watering system uses rainwater designed to move through the plants in the open wall. The open wall area is an opportunity for vertical green design and micro wind turbines to be added to the structure to help improve the air quality throughout the structure as well as the surrounding area. These design interventions become an asset to the site.

Image 3.23   Multiple-story parking structures are opportunities for green designs to improve the air quality due to the pollutants which exist from the the vehicles using the parking structure as well as from the vehicles on the road.

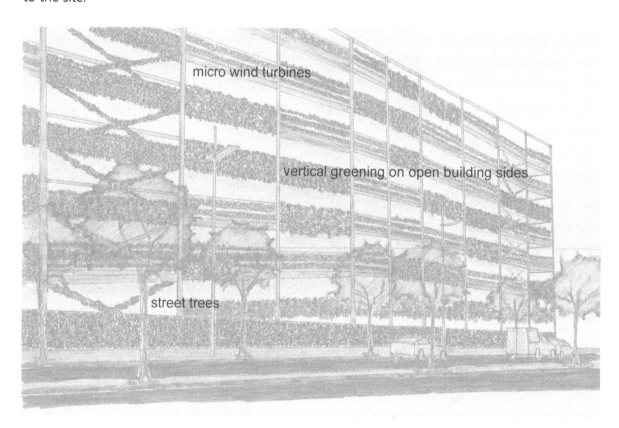

micro wind turbines

vertical greening on open building sides

street trees

Image 3.24   Labeled design of green parking garage exterior.

Image 3.25   A multi-story parking structure with a green design includes: street trees in the nearby sidewalks; micro wind turbines on the open vertical surfaces allow the wind to be used for increasing natural energy sources which can be used for lighting this building; vines and other creeping plants lace the open vertical structure improving air quality; LED signage indicates available open spaces and their locations so drivers can park quicker.

Creating the green design in Image 3.25 looks not only at the structure itself, it also includes the sidewalk and street around it. Street trees are added to create a layer of greening on the street level where people and drivers are located.

Outside the building there is a sign which indicates the quantity and location of parking spaces with an LED sign. The sign allows people to easily see if there are spaces while driving on the street and know where to access the open spaces. On the levels of the parking structure there are two main features: plants growing across the open structure and micro wind turbines. Using both in combination allows the parking structure to have multiple practices for a more effective sustainable design.

Micro wind turbines are added in some areas on the open walls. Micro wind turbines are built in multiple sizes and use the wind as a source

Image 3.26 This parking structure is ecologically designed with a surface of plants throughout the surface.

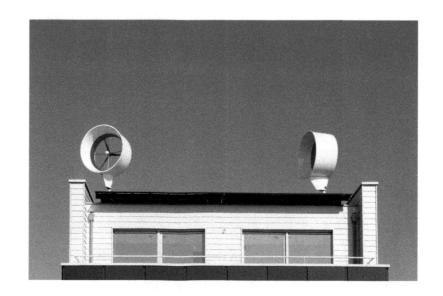

Image 3.27 Micro wind turbines on a residential building.

of energy. The small wind turbines are mounted on spaces where wind flows through. As the turbine blades turn a rotor captures the kinetic energy of the wind and converts it to electricity which can be stored in a battery source. The energy from the wind resource is meant to support the existing electric system, not be the sole provider of energy.

The green design in Image 3.25 uses a variety of plants which live and grow throughout a year. Planters are added throughout the levels of the parking structure which allow the growing plants to weave through the railings of the structure.

The exterior planters along the railing edge can also act as a water collector during a rain event. When this happens, rain runs along the railing structure allowing the plants to be watered naturally. The view of the building during the rain is mesmerizing with water drops weaving and dropping throughout the plants, creating a water feature along the building surface similar to that of a wire fence.

The goal of 10% of the surface area for greening is easily achieved and surpassed in this design of the exterior of the parking garage. As the interior and exterior of the building have a blurred edge due to the open sides of the building, the plants along the exterior also benefit the interior air quality. The next pages focus specifically on the interior greening.

Image 3.28   Parking garage with green design in Singapore.

Image 3.29   A multi-story parking structure with an entry sign showing quantity and locations of available parking spaces.

Inside the multiple-story parking structure the green design continues beyond the railing to planters throughout the parking levels. In Image 3.32, there is a green design in areas in lieu of some parking spaces as well as planters in corners and empty areas. Plants also grow up the columns and along the ceiling.

The plants used in this green design must be able to grow all year round as they are essentially in an outdoor site due to the open wall structure. The lighting of parking garages varies based on the distance to the open walls. However, there are LED lights throughout the parking structure for safety 24 hours, 7 days a week. Therefore the LED lights can help with the growth of the green design.

Image 3.30   The interior of a multiple-story parking structure.

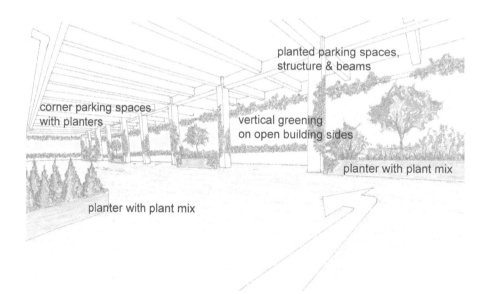

corner parking spaces with planters

planted parking spaces, structure & beams

vertical greening on open building sides

planter with plant mix

planter with plant mix

Image 3.31   Labeled design of green parking garage interior.

Image 3.32 The interior of a green parking structure includes: the exterior weaving of the vines and creeping plants are visible both from the exterior and interior; corner spaces which cannot be used for parking put toward greening spaces; parking spaces, structures, beams, and the ceiling have living designs weaving through the spaces; LED lighting above parking spaces in red or green indicates open parking spaces.

The green design in Image 3.32 uses available corner spaces where vehicles cannot be parked, as well as planted designs in other areas to help indicate driving paths. Some parking spaces have been changed to green spaces in order for there to be enough plants to improve the air quality with the emissions from the vehicles. Some parking spaces can have charging stations for electric cars supported by solar panels.

Image 3.33   Multiple story parking with living green design.

Image 3.34   Solar panels as solar canopy on top of parking garage. Solar panels convert sunlight into clean electricity and renewable energy and provide shade for cars.

The goal of 10% greening is achieved due to the combination of the plants on the open walls and rails of the building combined with the parking ramp design shown in Image 3.32. In parking structures it is important to have effective air circulation due to vehicle emissions. Another way to reduce the pollutants due to vehicle emissions is plants. There are plants which can reduce the pollutants in the air including emissions. As discussed in Chapter 1, the pollutants from vehicles are harmful for people to breathe into their lungs. The addition of plants including those which are specifically effective for reducing the pollutants from emissions can be used for the parking structure design.

Above the parking spaces are red and green lights to indicate available and unavailable parking spaces. Along with the signage on the outside of the parking structure which indicates the quantity of parking spaces and the levels they are on, this colorful system quickly allows drivers to see where there are open spaces without having to drive through aisle after aisle. One notable benefit of this lighting system is that the driver of the vehicle will not be slowly searching for a space and emitting more car emissions than necessary to find a parking spot.

Image 3.35   Red and green LED lights located on each floor above parking spaces visually indicate where there are spaces without having to drive around the levels.

The totality of this green design for both the interior and exterior greening is an asset for the health the physical and emotional health of numerous people. Drivers and employees of the parking garage benefit from the improved air quality as well as from enjoying the views and scents of the green design. The owners of the business get the benefit of using wind power to adjunct their energy use as well as show their support of a healthy green design to benefit the surrounding community with improved air quality. The community of people who live in the area ben-

Image 3.36   Parking spaces indicated by paver and patterns with plants growing through the open spaces. The planted spaces absorb water runoff.

efits from the improved air quality of the surrounding neighborhood. Additionally, the view to the parking garage is enjoyed by those walking by, driving by, or living nearby.

Image 3.37   Parking plaza for a train station and nearby area.

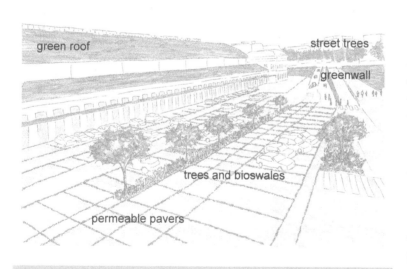

Image 3.38   Labeled green design for parking plaza.

Commercial spaces include the parking surrounding the shops or offices. Parking plazas are fantastic opportunities for greening spaces for a year-round planting design. In the site design in Image 3.39, there are permeable pavers, bioswales, parking lot trees, greenwalls, street trees, and green roofs on nearby buildings. Illustrated in the deep rich colors of Autumn, this parking plaza design has a pattern of plants which continue beyond the parking lot into nearby structures and further into the city. The goal of 10% of the space having green design is achieved successfully through the use of permeable pavers as well as removing some of the parking spaces for use as a bioswale.

Image 3.39 Outdoor parking plaza with green design: the parking lot includes permeable pavers, bioswales, and multiple trees and planted areas which add shade through the lot; the forested parking plaza is shown in Autumn; select parking spaces are repurposed for further plantings; the nearby building has a sloped green roof as well as glass surfaces which allows sunlight into the building for interior green design; the walls along the walkways having a living design, as well as trees added to the nearby plaza.

Permeable pavers come in a variety of patterns and sizes and allow water to permeate through the parking lot pavers to the water table. Images and further information about permeable pavers are found in the first chapter. Low-growing plants can easily grow in the spaces between the pavers and once established use rainwater, also known as stormwater, from the surrounding site for their long-term growth. Plants such as *Soleirolia soleirolii* – baby's tears, *Thymus serpyllum* – creeping thyme, *Isotoma fluviatilis* – blue star creeper,

Image 3.40   Vertical green wall in an urban plaza.

Sedum acre – goldmoss sedum, and *Sagina subrata* – Irish moss, can grow within the conditions of a parking lot where cars will leak fluids, metals from vehicles drop and decay, garbage is thrown, and people routinely walk. These plants tolerate the conditions and even better, they remediate the conditions of the parking lot air emissions, fluid emissions, and particulate matter emissions.

Within the parking lot in Image 3.39 there are also bioswales in addition to permeable pavers as part of the design. The bioswales have plants, shrubs, and trees which create needed shade in the lot. A bioswale is an area of a parking lot where water from the site flows to it by designing a slight slope throughout the ground toward the planted area. A parking lot can have multiple bioswales, all of which use the rainwater. Additionally, the rainwater moves the pollutants and particulate matter on the ground to the bioswale which allows the plants to filter and remediate the pollutants on the site.

GREEN STORM WATER INFRASTRUCTURE

Image 3.41 illustrates stormwater infrastructure for reuse of the water throughout multiple locations of a site. The benefits of rainwater reuse include naturally filtering water before it returns to the water table below ground. Reuse of stormwater on site also allows for the city or county waters filtering plant to not be overworked during times of storms as the water on a site is slowed down, reused, and filtered before leftover stormwater enters the storm drain toward the city filtering plant.

Image 3.41   Harvesting stormwater from rooftops, bioswales, and permeable pavers for irrigation before it is filtered underground is beneficial and a more sustainable long-term practice.

The layers of greening continue in Image 3.39 with a planted rooftop of the building, greenwalls along the plaza edges, and street trees added throughout the neighborhood to expand the greening to the extended community.

With the addition of planted roofs, the parking lot trees, street trees, bioswales, and the parking lot greenwalls, the area's green design reduces the heat island effect. Heat islands are areas,

Image 3.42   A bioswale on the sidewalk in Busan, South Korea.

typically in urban spaces, which experience higher temperatures than the nearby locations. This is because structures such as buildings, roads, parking lots, and other infrastructures absorb and re-emit the sun here more than planted landscapes and water bodies. The addition of the green design helps to reduce the heat island effects of typical large parking lots.

Image 3.43   Green parking lot at Wen-Xin Forest Park in Taichung, Taiwan.

With green design through the ground plane, eye plane, and overhead plane of the parking plaza and beyond in Image 3.39, consideration of a larger scope of greening in unique and achievable ways in this space is attainable. When planners, designers, committees, business owners, and communities are able to look at each parking plaza, each building, each walkway, and each individual space, they can develop green designs for their sites on small and larger scales.

Image 3.44   Mixed use urban space with greening throughout the space.

The greening of our urban and public spaces is an approach to combine green design as a functional, useful, alluring, and healthy setting as we find new ways to living with and working with the natural world. The designs in this chapter and book look at ways to explore a stronger relationship in the design of a variety of spaces with the natural world.

As public spaces vary in scale, use, and users, creating a design which celebrates a large range of users with different physical abilities, cultures, races, ages, income levels, and genders can attract people to a site, and even keep people longer in a site. A welcoming design is inclusive of people's health and comfort as well as the health of a larger picture of the future through sustainable design. The practice of using 10% greening as a goal can improve the short- and long-term health of a community with gathering spaces for people to engage and create even more green designs for their unique sites. The design of green, sustainable businesses and parking can become an asset to the community.

REFERENCES

Raanaas, R. K., Horgen Evensen, K., Rich, D., Siostrom, G., & Patil, G. (2011). Benefits of indoor plants on capacity in an office setting. *Journal of Environmental Psychology, 31*(1), 99–105.

Ulrich, R. S., Simons, R. F., Losito, B. D., Fiorito, L., Miles, M. A., & Zelson, M. (1991). Stress recovery during exposure to natural and urban environments. *Journal of Environmental Psychology, 11*(3), 201–230. https://www.sciencedirect.com/science/article/pii/S0272494405801847.

Time Invested Spaces

Lofts, Houses, Courtyards, Offices, and Classrooms

Image 4.1   Green designs for lofts, houses, courtyards, offices, and classrooms.

Anotable amount of time is spent in homes, offices, and classrooms, especially for those people who work from home, part or most of the time, or those who are homebound or caretakers, as well as people who work in an office or are students. There is an opportunity to design these time-invested spaces to be places of comfort, health, joy, support, and inspiration for those people who live, work, or study in these surroundings.

The spaces in this chapter include homes, a courtyard, an office, and a classroom. Homes vary in scale from studios and lofts to 1-bedroom walk-ups, to a shared flat or multi-bedroom condo with others, to a brownstone or house. In all these different styles there are open spaces for

DOI: 10.4324/9781003348634-4

light to shine through and magical green designs which can happen. Offices are diverse in their use and business type as well as from older private offices to modern open office designs. Classrooms vary from pre-school through tech schools, colleges, and universities, while also different in design based on the courses taught in room.

As discussed in the first chapter, green designs are beneficial for effecting the health of people and animals by improving air quality, emotional balance, and memory retention. The images in this chapter explore living designs toward these benefits. Green design is most effective when it defines the

Image 4.2   The illustration shows the initial design for the loft space with plants along the wall of windows as well as plants continuing on the shelf structure (that has no front or back) which can be viewed from the living room side or the bedroom side.

space while responding to the conditions and structures of the space. When exploring ecological design with homes of all scales, offices, and classrooms, both the interior and exterior, there are multiple approaches. It is effective to design with open spaces for people and animals to move, the ability to move large items, the options to change colors on some walls, and the ability to change art with changing tastes or to create a new surrounding.

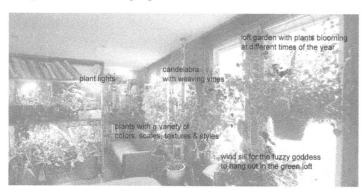

loft garden with plants blooming at different times of the year

candelabra with weaving vines

plant lights

plants with a variety of colors, scales, textures & styles

wind sill for the fuzzy goddess to hang out in the green loft

Image 4.3   The labeled photograph is taken from the opposite direction of the illustration to show a different perspective and fuller perspective from the bedroom of the loft space. The couch and bookshelves are an indicator of the space in both the illustration and the photograph. This photograph is taken two years after the illustrative design.

With this approach living green design can grow, move, and have unexpected magic throughout the weeks and years while enjoying your home. This means allowing for more living plant materials which by their very nature change often in magical ways. Green design spaces must be allowed to and encouraged to change frequently – just as people do. Some flexibility in green design is needed in a home, as well as some stability for long-term growth.

-

The initial design for the loft in Image 4.4 is merely to have plants. The initial design is not an interior greenspace as A. the greening did not define the

Image 4.4 A green loft design includes: a design that is friendly for the Goddess in the body of a cat to lounge, play, and eat; plants which grow at a relatively constant warm temperature; a design which blooms at different points of the year in a variety of colors, scales, and shapes; a design which grows in response to the light and structures of the space such as shelves, candelabra, and wires; plant lights on a timer for additional needed lights ; designing the green loft with a variety of scales of plants, loose flowing plants, plants with an upright structure, and plants with a variety of colors as well as a variety of shades of green; unexpected leaves and flowers which are revealed in a random day; the scent of fresh air; as the design overgrows the containers or the structure, plant cuttings are taken, propagated, and given to others; a water feature of a painting of an ocean adds a tranquil water focus; and the overarching design is on which allows a sense of encouraging unknowns to happen or chaos to grow to allow magic to be revealed.

space and B. the plants are movable and not designed into the space for long-term growth that responds to the space.

The first site in this chapter is being shown differently than the other design chapters. The illustration shows the site as it is proposed (the before image), and the after image is shown in a photograph two years after the site has been built and grown into the space, creating a green loft rather than plants as movable objects. This space is my artist loft.

The original drawing shown in Image 4.2 shows what the predicted design for the space would be as it grew, drawn with an estimated two years of healthy growth

Image 4.5 Interior plantscape using vertical green wall, terrariums, and hanging plants in Singapore.

after acquiring all the materials for the green loft design. The final design in Image 4.4 includes shelves which have attached plant lights, mature plants with large leaves, climber plants which rise and grow onto the exposed brick wall, and hanging plants which weave into structures such as candelabra and other wire-framed pieces. This creates a design which runs vertically as well as horizontally showcasing a variety of scales, colors, and textures.

The design accounts for ease of watering and allowing views from the windows to the courtyard and park across the street.

The majestic green loft design takes two years to develop and grow into the space and determine the lighting needs of the planted loft. The green loft design is a living green artwork which responds to the light and structures of the loft.

The design for the loft in Image 4.4 accounts for more than the artist that lives here, it also accounts for a playground and home for the fuzzy goddess, also known as a long-haired black cat. The green design must have cat safe materials as well as space for her to nap and play. The

Image 4.6 Industrial loft design with exposed brick wall and green vertical garden.

larger planters may have space with covered soil for her to nap and lounge. The leaves are also toys for playing with, while enough space is left on the windowsills for sunning, playing and petting.

The Goddess also enjoys walking through the green loft with a path from the windowsill through planters to the couch – a path she creates and changes at whim. This is relevant because it provides an example of the magic that comes with sharing a greening design with another and inspiring them to also make it their own. The unexpected changes which occur enhance the magical factor, further emphasizing the ephemeral qualities of time and space to add to the beauty of green design.

Image 4.7   Wall of planters in hanging style and shelving style.

There are over a dozen plant lights on a timer for 12 hours of light a day to increase the needed light for the green loft in Image 4.4. They are added to the space to allow the garden to grow in spaces that have low sunlight. For optimal growth opportunities, southern-facing windows naturally receive the most sunlight throughout the day and therefore the first place to look at when adding greening design options to an interior space. Unlike exterior spaces shown in previous chapters, designs within interior spaces are not influenced by varying temperatures. Interior designs are more influenced when the plants overgrow the space, need repotting, or new materials are acquired. When needed, the plants are trimmed and cuttings are given to others to create their own green lofts. Each year over 100 cutting from this loft are given to others who wish to have their green homes expand.

As there are no blinders or curtains on the windows, plants act not only aesthetically but functionally, creating partial blinders with an illustrative stained-glass artwork effect on the windows. Improved air quality and a fresh scent from the green design are additional benefits to living green artwork and design.

During the winter months a humidifier is in use frequently to keep the humidity at a comfortable level for the green design and those who live there. Maintenance on the green loft is minimal on a weekly basis with occasional days where overgrown areas of the design need specific care for structure or re-potting.

Some of the design in Image 4.4 is intentional, while other parts of it are unexpected magical accidents. The unexpected magical parts of the design are why I love designing spaces with living materials. Things change and grow in their own

Image 4.8   A room in a home with a relaxing living design on different layers, scales, and styles.

ways and at their own time. It is continually amazing to experience new flowers blooming unexpectedly, plants twisting around structures, new leaves, and seedlings sprouting from the soil.

Image 4.9   A brick wall along an open space of a loft with a large window is an opportunity to create a unique style green wall.

In this loft, plants are designed into the space, for the space, as an integral design of the total space, and grow through the space. Therefore the living design is not experienced as numerous separate objects within an interior, yet rather as living design which creates and responds to the space.

The effect of a green loft is stunningly important. Seeing the growth and change in the loft over time reminds people that things change, the worst minutes will grow into something better, the best minutes will be surpassed by ones which they cannot imagine, and the unexpected pieces of magic that come from seeing the green loft grow are countless. When a flower opens where months prior there was none, or new leaves are revealed, or the spaces of the loft are changing due to the living green art and design, the loft provides a space of inspiration during the easiest as well as the most challenging of times.

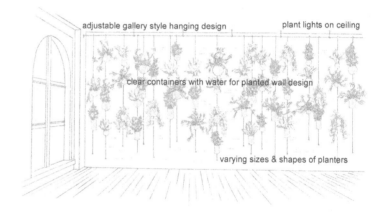

Image 4.10   Labeled design of green wall with living art.

Creating a living artwork or green design or green space is as unique as the space and people who use it. In Image 4.11, the exposed brick wall acts as a backdrop to a gallery style hanger system which uses clear jars to hold plants. The growing plant roots are visible through the clear glass. Having the roots of the plants visible serves multiple purposes. It allows the users within the space to see plants grow their own roots and people learn to make plant cuttings from these. The new plant cuttings can be put in water to take root and create even more plants, continuing the cycle to start again. People may relate to the idea of starting again and making new roots in their lives.

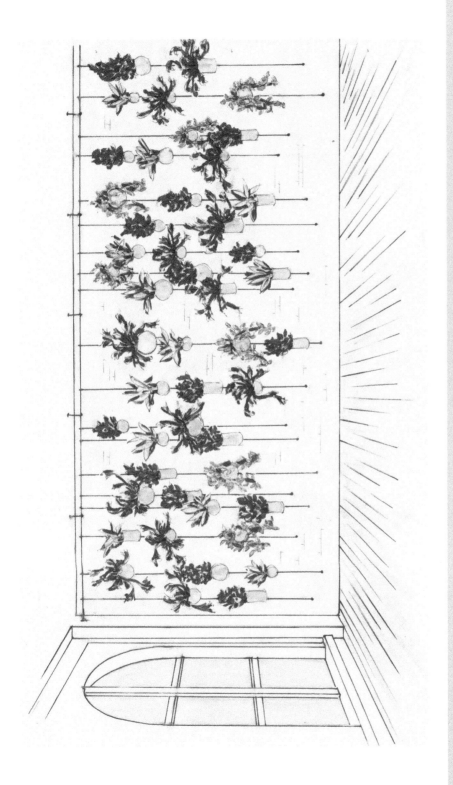

Image 4.11 A brick wall with living mural includes: gallery style hangers which have adjustable holders to be raised or lowered as well as move left or right; the clear containers allow the roots of plants to be visible; the water in the containers can be colored with food coloring to create a stained-glass effect; plants which can grow in water successfully include Friendship Bamboo (*Dracaena sanderiana*), Variegated Spiderwort (*Tradescantia zebrina*), Pothos (*Epipremnum aureum*), Spider Plant (*Shlorophytum comosum*) as well as others; when a plant grows beyond the scale of the container, a larger container can be substituted, or the plant can be split into two or more smaller containers, or the full plant can be placed in a pot with soil for further growth; and nutrients can be put in the water of the varying size and shaped containers for the plant's growth.

Artificial lighting is needed along the wall as the plants become too far away from the window for the needed sunlight as seen in Image 4.12. Adding timed plant lights on the ceiling along the wall will increase the full spectrum daylight needed for the plants to grow well. Timers for the lights are available for consistency for the light to turn on and off.

Having a full-length living wall can be created in different style spaces, even those with less light. The addition of plant lights or daylight bulbs directed toward the wall can make even rooms with small windows have a living wall. LED lights can be programmed as full spectrum lighting with the sunlight color of 5600–6000 degrees Kelvin. The lighting is not only beneficial for the plants, it is also easier on the eyes of users in the space.

Image 4.12   Diagram of the living wall showing different light needs. The light blue area next to the window will not need LED light for plant growth as the natural sunlight from the window is suitable for growth. The middle violet area will not get enough light from the window and will need LED lights on for a few hours for plant growth. The darker violet area will need more LED light for the plants to grow successfully.

Image 4.13   Some plants can be propagated in water from a cutting of an existing plant, and then grown in jars with water. As plants outgrow their containers, they can be placed in larger containers for continued growth or split into two or more smaller plants.

Using gallery style hanging, the brick wall design in Image 4.11 has plants' cuttings growing solely in water. The plants in the jars start as cuttings of larger plants if they are suitable to grow in water. In this method these cuttings are added to the jar filled with water, and the plant cutting develops roots. Once there are enough strong roots the cutting can be put into a pot with soil for long-term growth. Another option is to have a variety of jar sizes so that larger plants can continue to grow in water without soil. The hanging jars can also act as vases for cut flowers for the wall design.

The totality of the design is that there is an additional technique of living artwork which changes and grows in stunning ways while reaching beyond the wall as plants outgrowing the jars. Plants that have outgrown their spaces can be moved to other locations within the space, allowing for new plant material to be added to the green design in extended spaces. The 10% greening goal will take time in a design such as this and can be celebrated as the space takes roots.

Nutrients can be added to the water which is replenished as it evaporates into the air. There are natural nutrients for plants which can be made in a kitchen or provided with store-bought plant food. To create a stained-glass window effect, food coloring can be added to the water in the jars. The hanging jars can be any variety of sizes and shapes, adding to the diversity of the stained-glass effect.

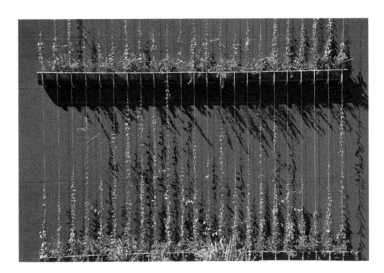

Image 4.14   Vertical garden in buildings located in São Paulo, Brazil.

Image 4.15   Hanging plants creating an ever-changing effect against a white wall.

Home offices are increasingly common as people continue to work in a hybrid format. A green office design can utilize prefabricated green structures as well as green structures that are site specifically built for your office. For the space in Image 4.18 there are a few notable pieces – the living desk, living wall artwork, an indoor tree, hanging plants, and smaller floor and desk plants. In a study, participants felt more peaceful and positive after spending 15 minutes in a room close to tall plants (about five feet away), compared to other objects which were further away (Han, 2019). Some plants also have a scent that is associated with energizing their mood, for focusing, or a scent for reducing stress. Inhaling lemon, mango, lavender, and other plant scents can alter gene activity and blood chemistry to reduce stress levels (American Chemical Society, 2009).

Image 4.16   A home office is common for those that have their own business or work from home. Creating a space that is inspiring while working from home is beneficial for the success of all involved.

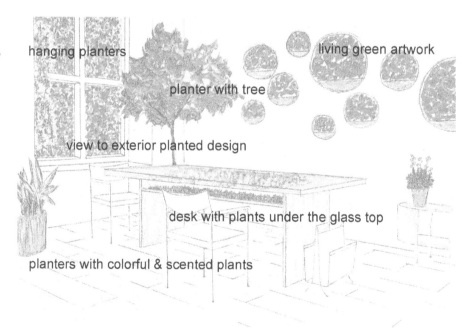

Image 4.17   Labeled green design of a home office.

Image 4.18   A home office with a green design includes: desk with plants growing under the glass top; living growing artwork on the walls; planters with a tree and other plants grow and change over time; hanging plants near window; scented plants such as mint can help with focus, other scents may be enjoyable; and a view to exterior plants extend the feeling of the open green space.

While sitting at the office desk in Image 4.18, there is healthy greening from every view. In this office design the desk is the centerpiece of the office. The desk itself has a glass top with moss, herbs, or short grasses growing underneath the glass top. Companies such as Habitat Horticulture make tables with a layer of living plants grown in and include a water catching structure. Designing your own is also an option for the creative builder.

On the wall there are framed planters showcasing living artwork and allo for meaningful observation as the plant material grows and changes over time. As with the table, there are multiple companies which make wall planters of different shapes and sizes, all with a way for the water to be caught within the frame itself. Gromeo and other companies make a variety of wall planters. Plant selection varies based on the lighting in the

Image 4.19   A living table created by Habitat Horticulture with an integrated water reservoir. There are prefabricated and custom designs of tables and other furniture.

Image 4.20   Living artwork can be bought from companies such as Gromeo, shown in the first image, or made for each site with the style, size, and type of plants that the office user enjoys.

room and the needs of the plant. Plants with similar needs of lighting, water, and soil type in the same planter create a stronger success for all the plants in a planter. Using a variety of sizes and shapes of the wall planters creates a different style green wall which acts as living art.

While working at the desk, there are hanging plants at the window which create a stunning interior foreground with a larger exterior garden view from the window in the background. The floor planter by the window with a large plant also adds to the foreground view. The larger exterior garden can appear to extend to the interior office through this design.

The indoor tree near the window in Image 4.18 is a small indoor tree. The tree can grow quickly in the first few years in the right conditions with light, water, nutrient soil, and a pot size which allows ample space for the roots to expand and grow.

Smaller plants are introduced to add pops of scent, shapes, and colors to other areas near the desk. Scented plants that can be grown indoors include jasmine, lavender, mint, begonia, heliotrope, hoya, eucalyptus, lemon balm, paperwhites, and a citrus tree. Smaller plants may be annual or perennial plants. Whereas annual plants have only one growing cycle, perennial plants continue to grow and reflower yearly. Annual plants can be enjoyed and then replaced as needed at the end of the growing period.

This office design in Image 4.18 meets the 10% greening goal with the wall planters, desk, hanging planters, and floor planters. The unique styles and forms of green design are ones which people can create in their own spaces to be reflective of their individual wants and needs for a well-lit and well-loved green design.

Image 4.21   The shelves have green ivy growing with LED lights built within the structures creating plant lights directly into the design.

Image 4.22   Shelving design with sunlight or artificial plant lights is an approach to green design which can be incorporated into the space.

Courtyards are exterior home spaces for people who live in larger buildings or complexes. In a chapter which looks at sites where people live, courtyards act as the semi-private front yard for all the people who live in the building.

The courtyard of the residential urban complex in Image 4.25 is the primary open green space that half of the building can view. The design includes permeable pavers, a planted edge along the entire building, window planters, a central open green area, additional planters, seating, and vines along the iron-framed entry. Across the street is a park with trees in Autumn colors.

Horizontal and vertical spaces are used to illustrate the green design opportunities which can be utilized to create a stunning urban courtyard. The open space needs to provide for diverse people to gather with seating and movable tables and large clustered areas of green design. The vertical greening and horizontal greening are for the views and use of the residents.

Image 4.23   Courtyard of a building in New York with over 300 people who live in the building and six floors of residents overlooking the space.

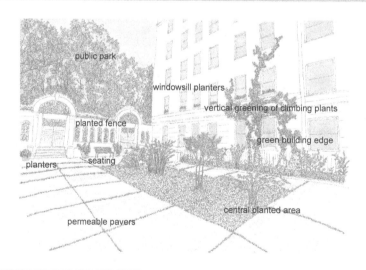

Image 4.24   Labeled design of a green courtyard in New York.

Image 4.25 A courtyard with a green design includes: permeable pavers across the site; large area of plants in the center of the courtyard with a depth of soil which can share soil for a richer growing condition; plants growing along the building edge in soil as well as vertically up the building; windowsill planters in each window; entry gate has plants growing through the metal structure; and the autumn scene feels expansive not only within the courtyard but also because of the visibility of the park across the street.

For those on higher floors, the views to the courtyard in Image 4.25 can be at a steep angle. Having vertical green on the windowsills and growing up the building gives more views to plants in the urban area. This design builds on ideas from the second chapter – designing for every resident to have a view to green space from their home. Vertical green walls help accomplish this. Creating a structure for plants to grow above the surface of the building is beneficial so that the building is not affected by the roots of the plants. This is currently done with installed, prefabricated exterior green walls.

Image 4.26   Home courtyard with open space for gatherings as well as for plants to grow together for years.

Within the courtyard, a continuous area of a planted design is more beneficial than single separated planters. This is for a number of intentions. First, smaller areas with separated containers do not allow the plants to grow together and create a healthy soil where the vegetation can grow larger due to the diversity of greenery. Additionally, the view of smaller separated plants for those overlooking the courtyard is not as appealing as larger areas of diverse plants growing together. Having green

Image 4.27   Plants in vertical design of home entry in Cordoba, Spain.

views can help improve people's physical and emotional health (Ulrich, 1991).

The courtyard space, while not meeting the goal of 10% greening goal, provides cleaner air and also psychological benefits for people that are able to view or experience the space. Throughout the year the space will have visible growth and changing color. The scent of fresh flowers and plants from the courtyard design is experienced by those on the lower floors as well as those who are in the space.

Image 4.25 is one courtyard of a multistory building on a block with other similar buildings. Looking at the larger scale of the block is explored in Chapter 2. Approaching green design on a larger scale of a neighborhood or block is an approach that works together with green design of one building at a time. When applying green design principles to a building, the interior and exterior of the site are designed one space at a time. Through each green design intervention in the urban fabric, a totality of greening adds up to a healthier community.

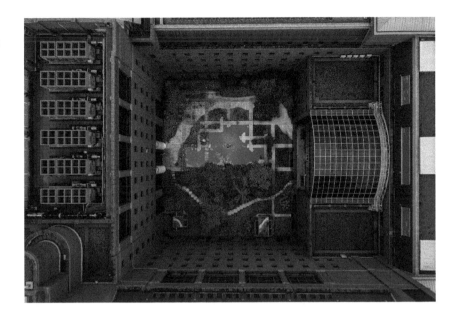

Image 4.28   An aerial view of a green courtyard in the center of an urban building, giving views and access to a lush garden design for people who live there.

Image 4.29   Street in Spello, Umbria, Italy, with plants throughout the space at different heights, colors, scales, and varieties create a lush comfortable space.

Indoor plants in an office can reduce fatigue during challenging work, more so than a five-minute break. Benefits of plants can also occur in offices with window views when there is a view to nature. These design methods are restorative approaches for offices and other spaces (Raanaas, 2011). Office settings have a range of styles and uses, generally with private spaces, semi-private spaces, open offices, and shared rooms. Image 4.32 illustrates a design of an office which celebrates living green design embedded throughout the space in unique approaches.

Image 4.30   Office gathering space and desks with large windows.

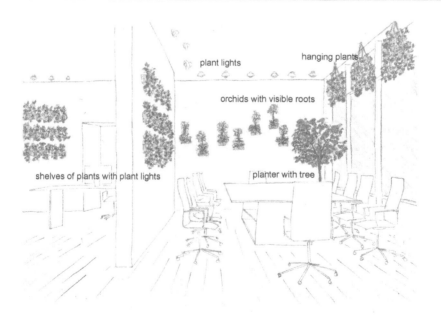

Image 4.31   Labeled green design for office meeting room and desks.

Image 4.32   An office meeting room with: shelves of plants with LED plant lights under each shelf on a timer used for lighting the plants under it; a wall with orchids which can grow without a pot or soil by being soaked in water on a routine schedule; LED plant lights on a timer, directed toward the wall of plants; a tree in the corner; and hanging plants with loose hanging leaves in front of the large windows.

The shared meeting room in Image 4.32 has plants vertically throughout the space. The living wall designs also weave throughout the office with plant lights added to the space to improve the growing success of the plants.

In this office caretaking of the greening can be combined with green team professionals as well as employees who want to learn more about plant care to jointly upkeep the living space. Companies can provide employee incentives as part of a program to save energy and have sustainable practices. Employees participating in the living green office design, alongside professionals, can help grow and expand long-term healthy practices.

The meeting room in this design is used for both employees and clients to gather and work. Having a setting where the green design is at eye level for those people sitting or standing helps the design to be more inclusive.

Image 4.33   Interior of Google office in Tel Aviv, Israel, with living design throughout the office.

Image 4.34   A Vanda Bitz's Heartthrob orchid comes in a variety of colors. They can be grown with bare roots. These are grown by soaking the roots in water on a routine schedule, then hanging the plant.

The green design in Image 4.32 includes an indoor tree, hanging plants, shelves of plants organically growing on the walls, and orchids which appear to float on the wall. The Vanda Bitz heartthrob orchids do not have to be planted in the orchid medium. These orchids can be designed to be placed on a structure or hung with exposed roots. The roots must be soaked in water which has plant food to continue growing and blooming.

Image 4.35 Living green design in an overhead structure with lights designed for the benefit of people and plants.

The shelving is designed with planters on the top side and plant lights on the lower side. On the top of the shelves there is a space for planting a series of plants and having a space for excess water to overflow and be safely held. On the lower side of each shelf there are LED lights which provide necessary light to the plants below for growth.

The office design has plants at different heights for people at varying eye lines, as well as greenery of different colors and blooming times. Having plants in an office, no matter the view from office windows, is an approach to design which engages people to be present and enjoy the ever-changing room.

Image 4.36 Living green shelves with a variety of plants growing in a shared space.

Classrooms are places for students to grow and expand their knowledge in formal and often traditional settings. Classrooms are changing to add technology; however, the addition of greening design can also work with technology and the learning process.

Image 4.37 A classroom with movable desks, a lecture stand, and a front board for projections or written information.

Classrooms are prime spaces for green design. Researchers from Cornell University found that nature can help people reduce stress. During finals week, professors at Cornell University decided to apply their research to a temporary design outside their library. Research shows that plants placed in a learning setting increase student attention capacity and realizations of studying (Ulrich, 1991). The same study found that interior plants can also prevent fatigue.

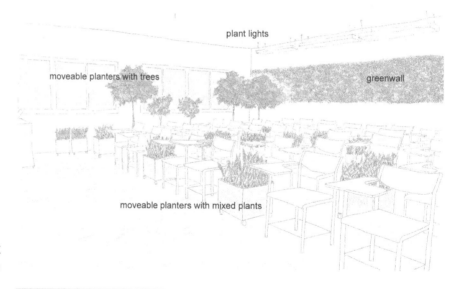

Image 4.38 Labeled design of a green classroom.

The design in Image 4.39 starts the learning process with greening within the classroom itself. The green classroom is a space for students to learn in healthy surroundings. These healthy surroundings create a peaceful and inviting setting which may inspire people throughout their careers.

Image 4.39   A green classroom design with: green wall on the back wall which can be watered with grey water from the building. Plant lights on the ceiling on the back wall which change throughout a day to replicate sunrise to sunset through technology. Peace Lilies as well as other plants and trees are in movable planters to provide flexibility in different classes. Low-height plants are in the planters between seats to allow clear visibility to the front of the classroom. The movable planters are built with an area to catch any water overflow. Students in a horticulture program can learn hands-on knowledge by maintaining the classroom living design. Fresh air is an effective design in the room throughout the year which is a healthy and lovely space for all those that use the room.

In the classroom green design in Image 4.39 there are movable planters designed between the seating with plants that are tall enough to see from any seat, while also below eye level to not block views to the front of the room. The planters are all on wheels for each moving as well as have a space designed into the planter to catch any water if it was over-watered.

The plants within the chair space are low light to medium light plants, such as peace lilies. As classrooms are not used and may have the lights turned off on weekends, selection of a suitable low to medium light plant is needed.

The movable planters with indoor trees are along the side of the classroom to add height to the green design as well as have plants which show more growth over time. With enough time in the classroom, students and faculty can see significant growth in trees during their learning process of a semester or school years. As the trees grow beyond the size of the planter, the trees can be replanted in larger sites such as a community park.

The goal of 10% of the surface space having green design is achieved in this design with the majority of the greening being located on the back green wall which accounts for no greening on the front classroom wall. The effects of fresh air, views to growing designs, and changing colors are benefits for all users of the space.

Image 4.40 Trees in mobile planters at Whitney Museum.

Image 4.41 Night view of the installation with a few weeks of growth with plant lights. When the trees outgrew the planters, they were planted outside at the end of the installation.

Along the back wall in Image 4.39 there is a large green wall and plant lights on the ceiling. The large green wall has soil for the plants as well as an irrigation system of grey water from the building used for the large wall planter.

Natural sunlight and sunlight bulbs help plants have greater longevity. Plant lights or full-spectrum lightbulbs are full-spectrum lights or in other terms, white sunlight. There are also LED lights which can be used as plants lights. Through the use of technology, LED lights can vary in color from sunlight to pink or green for example. Changing light colors can be used with an app connected to the lights. The app can turn the lights on in the morning with the lights appearing to be warm morning sunlight. Throughout the day the plant lights then change to white light and then again back to warm sunset lighting. This mimics the natural light outdoors and can be helpful for plant growth as it mimics natural sunlight.

As this is a space of education, students as interns in landscape design, botany,

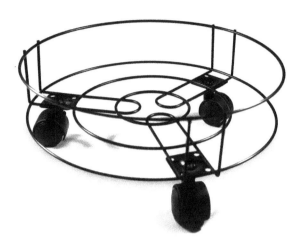

Image 4.42   Base with wheels for planters. Bases are made in a variety of sizes and shapes to fit planters.

Image 4.43   Hanging plants and vegetables growing on a wall structure.

horticulture, or other studies can be in charge of maintaining the site and addressing the changing needs as it grows. Plants throughout this classroom can be colorful or vary in shades of green from yellow greens, blue greens, silvery greens, and dark greens. The shapes of the plant leaves can be similar or vary greatly. It is the totality of the green design in the classroom that makes the classroom a highly successful space for the retention of knowledge due to the views to living plants, breathing of fresh air, and changes of the spaces due to plant growth, throughout the weeks, months, and years of the school.

Image 4.44   Loft with illustrious garden weaving through the space.

Seasonal affective disorder, also known as S.A.D, is a type of depression that takes place starting in the Autumn due to the decrease in the hours of daylight. Adding daylight bulbs and LED lights to a space is not only beneficial for plant health, it can also be beneficial for people. Increasing the hours and quality of interior daylight that people are exposed to in the autumn and winter, where there are decreased hours of outdoor sunlight, can lessen the effects of S.A.D for some people.

Additionally, as shown by multiple studies (Han, 2019; Nuccitelli, 2023; Ulrich, 1991; Wells, 2003; WHO, 2016; Wolf, 2010), exposure to green design is beneficial for people's health for stress reduction and healing. Caring for a garden and green design is also shown to reduce stress and improve physical and emotional health. As homes can be used by a variety of different users with different needs, exploring all the needs of a family, from a kitchen garden with herbs and edible plants, to green walls for offices, to green design for communities strengthens the health of both small and larger sites.

The totality of the green design of spaces with fresh scent and continually growing plants is integral for people, animals, and the larger topics of a healthy earth. With the 10% goal of greening as a new approach to interior, exterior, private, and public spaces, the results can have overwhelmingly beneficial impacts on a small and larger scale. The designs in *Ways of Greening* create the feeling of comfort and peace by spending invested time in green spaces. In the sites in this chapter, bringing a green solace to everyday spaces is achieved through unique designs.

REFERENCES

American Chemical Society. (2009, July 23). Stop and smell the flowers – The scent really can soothe stress. *ScienceDaily*. www.sciencedaily.com/releases/2009/07/090722110901.htm. (retrieved 26 April 2023).

Han, K.-T., & Ruan, L.-W. (2019). Effects of indoor plants on self-reported perceptions: A systemic review. *Sustainability*, *11*(16), 4506. http://doi.org/10.3390/su11164506.

Nuccitelli, D. (2023). *The little-known physical and mental health benefits of urban trees*. Yale Climate Connections. https://yaleclimateconnections.org/2023/02/the-little-known-physical-and-mental-health -benefits-of-urban-trees/. (retrieved 12 March 2023).

Raanaas, R. K., Horgen Evensen, K., Rich, D., Siostrom, G., & Patil, G. (2011). Benefits of indoor plants on capacity in an office setting. *Journal of Environmental Psychology*, *31*(1), 99–105.

Ulrich, R. S., Simons, R. F., Losito, B. D., Fiorito, L., Miles, M. A., & Zelson, M. (1991). Stress recovery during exposure to natural and urban environments. *Journal of Environmental Psychology*, *11*(3), 201–230. https://www.sciencedirect.com/science/article/pii/S0272494405801847.

Wells, N., & Evans, G. (2003). A room with a view helps rural children deal with life's stresses. *Environment and Behavior*, *35*(3), 311–330.

Wolf, K. L., & Flora, K. (2010). Mental health and function-A literature review. In *Green cities: Good health*. College of the Environment, University of Washington. www.greenhealth.washington.edu.

World Health Organization. (2016). *Urban green spaces and health*. Geneva: WHO Regional Office of Europe.

Chapter 5

Private and Public Spaces

Image 5.1   Green design for warehouse park, older architectural space, spiritual center, courthouse steps, and Penn Station.

This chapter explores privately owned spaces that are commonly used by the public. Though not public-owned, these are spaces where people gather and use routinely. These spaces affect community design and community identity and can be used as spaces for improving the emotional health for its members.

Businesses which serve the community can reap the benefits of a green design for the employees, visitors, and larger community. There are benefits which can be seen in employee retention as well as users who continue to return to a business to experience a growing green design. With the types of businesses in this chapter, parks, spiritual centers, legal services, travel, and shopping, consumers have multiple choices in their area. Consumers have multiple options to

DOI: 10.4324/9781003348634-5

choose from among their community businesses; therefore creating a site which improves the health of the people and space is an asset for loyal consumers. To have a consumer choose a site as one for repeat use means they have found qualities in the business and owners worth returning to, such as a green healthy site.

Health is not a trend and rather is necessary as a long-term approach and lifestyle for a healthy community. The decision to employ green design in a business for the benefit of the customer and employees' physical and emotional health can add to the legacy of the business. The 10% greening goal is a long-term approach to a healthy site and community design, which can lead to people who invest in their community.

Large open spaces such as a warehouse are opportunities for park spaces which can be enjoyed all year round in places with colder weather. The design in Image 5.6 creates a park with multiple changing uses of the green space.

This indoor park has herbs and vegetables, as well as trees, shrubs, grasses, and plants. An irrigation system suitable for each area of

Image 5.2 An empty warehouse with large windows throughout the space becomes an indoor park with year-round use.

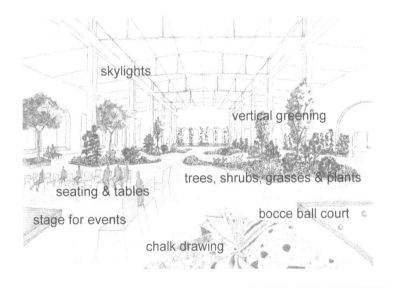

Image 5.3 Labeled green design of an interior park in a warehouse.

plants is a visible part of the design, allowing people to see the watering process. There is also a water feature for a soothing sound in the space. Movable benches, chairs, and tables are for reading, talking, board games such as chess. There is also a stage for people to have theatrical or musical events.

With the exception of the stage and planted areas, other areas of the warehouse park, Image 5.6, can have flexibility in their use. Some examples of flexible use areas include a piano, bocce ball court, chalk drawings on the floor, floor-size chess or checkers board game, movable seating and tables, lighting, electric sources throughout the space, and internet access.

Image 5.4   Greenhouse that is also a park with a lush botanical garden for visitors to enjoy all year round.

To have the plants grow more easily in the warehouse park, skylights have been added throughout the space. LED lights for night use or extended light throughout the winter and days of short light are also added. Spending time in these lights as well as the entire park can help reduce the effects of S.A.D. (Seasonal Affective Disorder). Stained-glass artwork can be added to the windows for a dramatic effect as sunlight streams through the windows creating colorful artworks cascading into the space.

Image 5.5   Green wall garden in an ecofriendly design for a building interior.

The goal of 10% greening may fall slightly short in the current design to allow other uses for the park in this space. It is at least 8% greening of the surface areas of the warehouse if not 10%. Further green design can be utilized in the larger quantities in other rooms of the warehouse which may help meet the 10% greening goal of this large space.

Image 5.6 An indoor park in the warehouse with added trees, shrubs, grasses, flowering plants, herbs, vegetables, and other edible plants throughout the space, both on horizontal surfaces and vertical surfaces. Other park activities include a bocce ball court, piano, stage and seating for musical and theatrical events; seating and tables for talking, reading, chess, and other ways to enjoy the space for an extended time. Changing chalk drawing by a local artist on the floor can also be used for a floor scale chess game; irrigation and a water feature from rooftop rainwater; skylights and large windows for sunlight, with added stained-glass art; LED lights for night use of the park are also included in this design.

Adding a green design to the exterior of the warehouse, Image 5.6, to both the rooftop and walls, can extend the space of the park beyond the interior design. This extension of the design can weave into other sites throughout the city.

The warehouse park is inspired by other interior and exterior green designs, much of which is shown throughout this book. Further examples not shown in this book include the Ford Foundation Garden in New York City, as well as an indoor pop-up park titled "Park Here" at the Openhouse Gallery in New York City from late 2010 to early 2011. Numerous green atriums, rooftop agricultural gardens, pop-up parks, hydroponic gardens for restaurants both interior and on rooftops, the Bentway Park in Toronto, Canada, under a highway, multiple story green walls both interior and exterior, and numerous other spectacular green designs serve as inspiration for unique green design.

Image 5.7   This indoor garden near a pond is an example of creating an interior space with a lush green experience.

Image 5.8   Residential buildings with lush green roofs and surrounding green design in Chengdu, Sichuan, China.

Reimagining greening for existing spaces means reinvigorating the space with a different and new life while celebrating the past meaning of the space. In the site on Image 5.13 the skylight acts as a central focus in the space for stunning attention, as well as natural light for a tree.

Adding a magnificent blue Chinese wisteria (*Wisteria sinensis*) to the interior brings a sense of growth and future to a room which is visibly breaking down and aging. The contrast of the deteriorating materials in the architectural style of a long-ago era, alongside the impressive tree of blue and green foliage, creates a striking design with a sense of a past and future both in the same space.

Image 5.9   The skylight in this space between the columns creates an area of central focus for a green design.

illustrious tree under skylight

Image 5.10   Labeled green design of interior site with skylight.

Though the goal of 10% greening is not met in the design of Image 5.13, with the showiness and scale of the tree, the space becomes a greening room where people can sit, read, draw, contemplate, gather, talk, or be inspired. With the addition of the central focus of the blue Chinese Wisteria, people who visit the site can experience a long-ago past in a space as well as a long-ahead future for it.

With the design choice to add one notable tree which has sculptural growth, a light fresh scent and changing colors of foliage, the space dramatically changes. It is an inspiring design which illustrates the past and the future so clearly shown in one space. While sitting on one of the benches in the space people may want to consider their roles in times of the past and open possibilities for the future.

The solid Tuscan style columns in the architecture are contrasted by the sculptural and weaving form of the blue Chinese Wisteria. The skylight creates a halo effect directed down onto the colorful tree, illuminating it with natural sunlight.

Image 5.11   A greenhouse with trees growing throughout the space.

Image 5.12   Tree roots inside of an ancient wall in Ayutthaya, Thailand.

Existing spaces, particularly those with distinctive styles of architecture and use, require people with fresh eyes and perspectives to find green design solutions. Existing spaces can, at times, be overlooked for sustainable design alterations. This may be because some people can only see the spaces for what they presently are rather than for what they can be. To help reconceive spaces for what they can be is something every artist and designer has their own technique for.

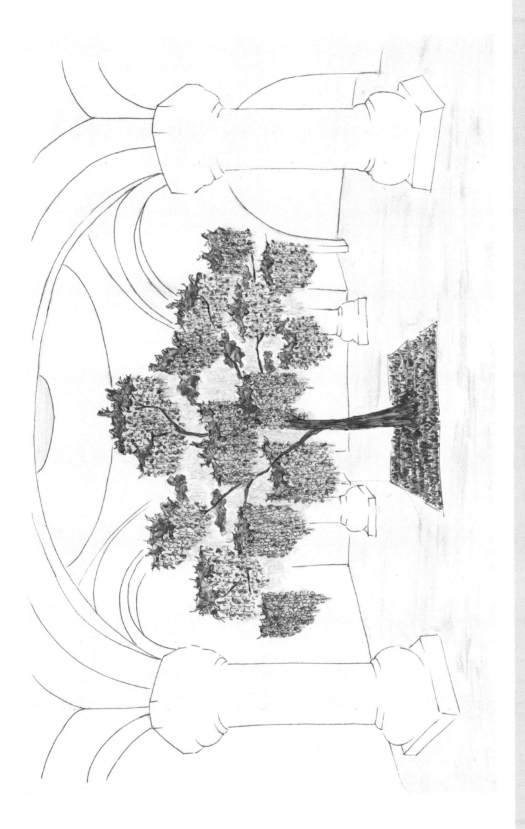

Image 5.13   A Blue Chinese Wisteria (*Wisteria sinensis*) under the skylight for a tree to grow toward the natural sunlight; a large planter above and below the floor for the tree to have space for the roots to grow; and add seating to space for people to engage in the space for an extended time comfortably.

It is important to make sure that not only newly built spaces are utilized for green design, but also older spaces are also used for green revitalization. With exception, spaces with a past are permitted to change, reflecting the past as well as the present and future. Due to the differences of each site, sustainable design with living plant materials is unique to each location. This uniqueness is a valuable attribute which can bring people to the same site multiple times to experience, see change, and be in the space for an extended time.

Spiritual centers, no matter their religion or practice, have a sense of power in their architectural design, colors, materials, acoustics, views, lighting, steps, heights, and scale in the full scope of the space. Spiritual centers encourage visitors to look at the larger universe as well as to celebrate their individual roles and actions for peace within themselves and the larger universe.

Areas and spiritual centers vary in scale, including areas for larger

Image 5.14   The ancient ruins and tree roots of a historic Khmer temple in the temple complex of Angkor Wat in Cambodia.

Image 5.15   Ta Som Temple in Angkor Wat, Cambodia, with tree roots on Stone Buildings creating a dramatic entry.

congregations and gatherings as well as smaller groups and private dialogues in some areas. Adding green design and gardens to spiritual centers is a beautiful way to celebrate growth, time, and change with fresh air and blooming throughout the entire year.

In the design in Image 5.20 of a semi-private space within the spiritual center, there is seating throughout, inviting people to sit alone or in small groups. Comfortable seating within the green design is important as it is intended to encourage people to stay longer with time for their

Image 5.16 Spiritual centers can be loftier, peaceful, and tranquil spaces with high ceilings, altars, seating, private spaces, and dramatic lighting which are designed to help visitors understand a bigger idea while having inner explorations.

own thoughts and healing. Visitors can breathe in fresh air while contemplating their priorities, inner thoughts, and spirituality.

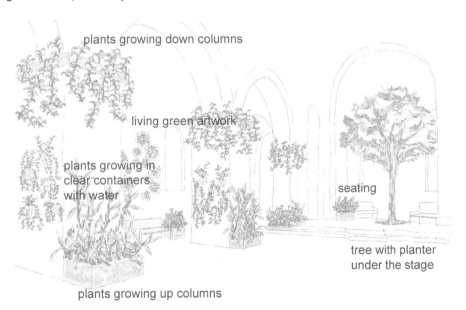

plants growing down columns

living green artwork

plants growing in clear containers with water

seating

tree with planter under the stage

plants growing up columns

Image 5.17 Labeled green spiritual center.

In this site the green design has planters on the ground with structured plants growing upward. Above people's heads there are also planters with plants which hang and grow loosely. Between these two planters are plants grown in water, similar to the brick wall design in Chapter 4. In the site Image 5.20, the plants grown in water can be cuttings, flowers, or full-size plants with varying size jars.

The contrast of the structured plants growing upright, the loose plants flowing downward, and the variety of plants growing in water can help people to see multiple choices in design as well as in their own options for their lives. While sitting in this space, the great range of plant styles, structures, colors, scale, and scent can inspire people of the great gamut of choices in their own current experiences and options.

Image 5.18   Some plants can be propagated in water from a cutting of an existing plant, and then grown in jars with water. As plants outgrow their containers, they can be placed in larger containers for continued growth or split into two or more smaller plants.

A water feature of small scale can be located in different areas of the room which adds soothing sounds that are relaxing for the users of the space.

On the walls there are structures for living art. As with Chapter 4, these living artworks can be bought or built for the site. Through their experience in the diverse green space with living artworks, visitors can be reminded of a great variety of styles and choices in their own to opportunities.

Image 5.19   Plants with a loose form grown throughout this site can include a lemon lime pothos, variegated aglaonema (pink and green), peace lily.

Image 5.20   A green design for a spiritual center includes: creating a semi-private space where people can listen to their thoughts or have semi-private conversations with fresh air indoors; multiple comfortable seats are placed throughout the site for contemplation; live artwork on the walls, framed and draping downward and growing upward; plants growing in jars with water (which can be colorful water) have roots which are visible through the clear jars; water feature for a soothing sound and a large sculpturally stunning tree is a focus of the design with roots growing in a structure below the stage.

The tree on the raised stage is a focal point of the design in Image 5.20. The tree is a showy tree which grows indoors with a structure built under the stage for soil and support. Having one large attractive tree can be effective whether it is a raised area, lower area, or same level as the rest of the floor. Around the tree there is seating to enjoy the view not only of the tree but the entire space from a slightly higher viewpoint. Through the use of the large tree along with vertical and horizontal greening throughout the space, the 10% greening goal is achieved in this space.

Image 5.21   Icelandic moss in living interior wall art.

Creating a natural space in the spiritual center can be a unique asset to the center to attract users to explore spirituality while looking at natural design. Being in a natural design with growing plants, fresh air, and scented greenery is a goal of the green design within the spiritual center or place or worship. Green design installations improve healing and wellness and reduce stress (Han, 2019; Nuccitelli, 2023; Ulrich, 1991; Wells, 2003; WHO, 2016; Wolf, 2010). Those caring for and maintaining a green design such as this may also experience the relaxing and healing benefits which are associated with working with plants.

Image 5.22   Tree growing through the floor of an abandoned factory with green moss.

Courthouses have a powerful architecture design. Courthouses support a myriad of services – from court cases of felonies, marriage certificates and ceremonies, and a full range of other legislative purposes. The buildings are designed to look formidable – steps require people to walk up to the higher ground and authority of the court. The columns are oversized and tower over people, appearing to rise to the sky. There are statues and quotes along the building which create an entry that is towering over the height of an individual. These are not designed to be comfortable spaces.

The entire design is towering and powerful and can be intimidating to people. If the role of the design is meant to be welcoming and comfortable, the design falls drastically short. To design for comfort, creating spaces of the scale of an individual, these unique court gardens and green spaces have multiple qualities.

The design in Image 5.27 has fresh scents, plants occasionally blowing the wind, and space to sit and pause

Image 5.23   With courthouses having a range of purposes, creating and retrofitting them to feel authoritative while also being of the people and for the people – all the people – can help make these buildings feel more approachable and have spaces of comfort for those that need it within the justice system.

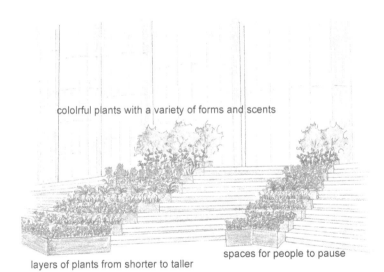

cololrful plants with a variety of forms and scents

layers of plants from shorter to taller

spaces for people to pause

Image 5.24   Labeled design of courthouse with greening.

for a minute or an extended time for anyone that needs it. These courthouse green designs create spaces for pause, reflection, or simply to gather thoughts before, during, or after their court appearance. The green design proposed in the illustration takes the scale of the entry steps down to pedestrian scale.

Making the green design entry feel welcoming in Image 5.27, adding handrails with the planted areas, and at a pedestrian scale is important because it takes into account that the court system is to help people – all people. Through the green design, the entry is re-scaled at the steps to have a pedestrian space for people to pause and reflect.

The design in Image 5.27 is beneficial for people who are at the courthouse daily or on occasion. People who work at a courthouse can sit on the steps, gather their thoughts, talk, share time, and remember that they work there to support people and the process of the court system. Additionally, the design is beneficial for those who go to a courthouse infrequently for several of the same intentions noted above.

The green design helps create contrast against the tall structured columns with flexible plants that sway in the wind. This green design gives a space for users of the court to pause and consider their role in helping people, culture, and society.

The planters have plants which are shorter in height at the lower steps, to plants which are taller at the highest step, with a range of plant heights in between the two. The goal of greening 10% of the surface space is not achieved in this design. The success of this design addresses the visual re-scaling of the architecture to be balanced for people.

Additionally, interior green design of a courthouse can help add a pedestrian scale and provide additional and more comfortable spaces to pause and breathe throughout the court processes.

Image 5.25 The Capuchin garden, garden of the courthouse in Chateauroux, France, creates a pedestrian entry with a friendly scale.

Image 5.26 Courthouse in Santa Barbara, California, with a green design to the side of the entry steps.

The entry and the exit are the same for the courthouse in Image 5.27, and therefore considering the experience of entry and exit for the users is important. To start and end a day passing green design which sways in the wind,

Image 5.27   A courthouse with pedestrian scale green design at entry includes: a place for people to stop and pause and sit on steps with colorful planted views; handrails added along the planters for ease of walking; layers of plants ranging from a height of a few inches through the height of a body; plants can have upright structure or looser structures which move with the wind, have scent and bloom at different times of the year; and the green design can extend into the courthouse corridors and rooms.

has colorful growing plants, fresh scent, and a space to pause provides benefits to all users of the site.

I live across the street from a courthouse in the Bronx, New York. It has an oversized entry on all four sides of the site, with layers of steps which I use for running drills, as well as to sit at the top of at night to look out on an expansive view of my neighborhood. It is part of my neighborhood, and there are people that hang out on the steps at different times of the day and night. Creating a green space within urban areas requires unique thinking, and this is one of those unique spaces that can benefit the neighborhood.

Image 5.27 is not just about adding plants or even gardens, it is about green spatial design and reconsidering how we engage with the spaces we inhabit. Whether we inhabit the space for a quick minute while walking through, or for an extended time while staying on a frequent or infrequent basis, designing spaces for comfort, ease, health, and change over time is the goal. Green design can help ease some of the discomfort and scale of courthouses. This contrast of the green design to the oversized unmovable structure is important for some reflection and change for people.

Image 5.28 Park in front of the courthouse in Hamburg.

Image 5.29 Administrative building with park in the winter.

Penn Station is a central hub for travelers to New York City. This space has countless amounts of moving and ever-changing people, all year round and during all parts of the day and night. People are on site for a few minutes or few hours as their trains run smoothly or are delayed. Commuters are on site routinely while other people may be at Penn Station less frequently.

The design in Image 5.34 starts by adding much-needed and surprisingly missing seating to the space which allows people to sit on seats in lieu of the floor, which is their current option. More notably, the addition of trees in mobile planters makes the spaces have healthy green structures which grow, bloom, and change over time as people visit the Penn Station on their way through to their destinations.

In Image 5.34, mobile planters have wheels which can be locked to stay in place or unlocked to easily roll. Using mobile planters allows the plants to be

Image 5.30   Penn Station in New York City is a space where trains run all day and night. The recently renovated design is sunny and open and an opportunity for green design.

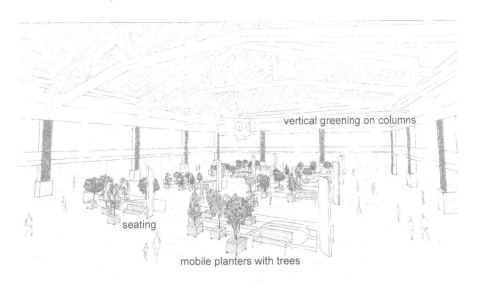

Image 5.31   Labeled green design of Penn Station, New York.

easily moved as needed if there is an occasion to do so, while also allowing clusters of trees to be placed and grown together or moved into evenly spaced designs.

As noted in the first chapter, the addition of greening in a space can help reduce sound waves. Sounds echo in the large open space of Penn Station including mechanical sounds, sounds of people talking, walking, and rolling luggage. These sounds are reduced when there are soft surfaces or pockets of soft spaces for the sound waves to be decreased. The denser the greening throughout Penn Station, the more the quantity of soft surfaces and soft pockets the sound waves can be absorbed and reduced.

Along with reduction of sound, there are psychological benefits of walking through an urban green oasis during all times of the year. For people who work at or are routinely at Penn Station, the benefits of being in a planted space include reduced stress, reduced A.D.D (Attention Deficit Disorder) in children, as well as provide physical activity, calming and inspiring surroundings, improved cognitive functions and emotional growth (Han, 2019; Nuccitelli, 2023; Ulrich, 1991, Wolf, 2010).

As the trees grow larger than the size of planter, the trees can be re-planted in long-term sites such as street trees, in arboretums, lobbies, offices, homes, or other places which are suitable for the tree to grow to

Image 5.32 Opened in 2019, the Jewel Changi Airport is a mixed-use space with commercial spaces in Singapore.

Image 5.33 Interior of Changi Airport in Singapore, Southeast Asia.

Image 5.34 Penn Station with green design includes: add seating for people waiting for their trains or friends (presently there is no seating and people can be seen sitting on the floor); green design while with mobile planters while keeping all signage of trains visible; structures and columns around the central area have plants growing upward and downward; mobile planters with a variety of trees grow throughout the space; when the trees grow too large for the planter they can be planted outside; and the green design makes the site pedestrian scale and has sub-spaces for people to pause in the space while waiting for their train.

its full scale and life. Using greening in public sites in unique ways can also inspire people to be able to apply greening to their own private sites.

One other notable design addition is seating. At Penn Station there are people waiting for a friend, colleague, or family member as trains arrive. Other people arrive early for their train, while others shop at Penn Station. Seating is available on the lower level near the trains, not on the ground level of shops, signs, entries, exits, and more. It is beneficial for people of all ages and abilities to have seating in public spaces. Seating allows people to spend some time pausing, talking to a friend, organizing their items or thoughts, and enjoying the space with ever-growing urban greening.

Image 5.35 Sustainable architecture for a residential building in Turin, Italy.

The combination of green design and seating in a public space is a simple yet highly effective approach to having Penn Station be healthier, green, more comfortable, beautiful, and enjoyable with its ever-changing scenes.

In urban areas it is unlikely that buildings will be torn down to make new public green spaces. Therefore, creatively greening existing structures and sites is the approach this green design specialist and author takes throughout this book.

Image 5.36 Tropical indoor garden in Hamad International Airport, Doha, Qatar.

Image 5.37   An artist's urban loft with a vertical green design.

In interior and exterior sites, from small-scale through large-scale settings, greening interventions can be unique assets. Designing can be from the floor up, from the sky down, from larger scale to details, or starting from the details, or most likely a combination of all of these. Residentials, visitors, business owners, employees and community members all become a part of a greener healthy lifestyle as well as participants in the site design and process. People can feel invested in their own green community and can choose to participate in the growing process of different sites and at various stages of care. Through people participating in the process of greening, communities can grow stronger through the shared practice of greening.

By adding at least 10% surface feet of space with a living design the air humidity is more comfortable for breathing. Additionally, the views to the living design that changes, grows, and blooms encourage people to enjoy the process and details and see change in visible time.

There is a beauty and ease of living and working in spaces which combine urban communities with a seamless infrastructure of weaved living designs. The psychological benefits, physical benefits, and economic expansions of a healthy green site and community have short- and long-term successes and sustainability. The approach to green design illustrated throughout this book encourages people to uniquely design these types of spaces as well as a myriad more.

Designing with living plants as a primary material is something everyone can do.

Image 5.38   Urban green design, designing from the ground upward, and the roof downward, as well as big picture greening and small details of green magic in the design.

The scale and layers of the designs may require multiple people and professionals. However, at some scale, everyone can participate in adding planted designs to improve air quality and reduce stress at their home or work surroundings.

These types of living designs are both big picture of time and change while also small details of time and change. The contents of this book are to help people re-see, reimagine, and reconceive how easy green designs can be applied to spaces for

-improved diversity, equity, and inclusion (DEI) in communities,
-improved learning settings,
-improved office settings,
-improved living surroundings,
-improved psychological and physical health,
-and for improved sustainable practices.

These types of interventions in design cannot be understated.

It is with all these images and words that this author and green design specialist believes in your success to live these pages out in designing your sites. Everything is possible – look at how far we have come throughout this book and with green designs as a whole. Let yourself green spaces into reality for yourself and people beyond you. You can go as far as you let yourself – which is infinite!

REFERENCES

Han, K.-T., & Ruan, L.-W. (2019). Effects of indoor plants on self-reported perceptions: A systemic review. *Sustainability, 11*(16), 4506. http://doi.org/10.3390/su11164506.

Nuccitelli, D. (2023). *The little-known physical and mental health benefits of urban trees.* Yale Climate Connections. https://yaleclimateconnections.org/2023/02/the-little-known-physical-and-mental-health-benefits-of-urban-trees/. (retrieved 12 March 2023).

Ulrich, R. S., Simons, R. F., Losito, B. D., Fiorito, L., Miles, M. A., & Zelson, M. (1991). Stress recovery during exposure to natural and urban environments. *Journal of Environmental Psychology, 11*(3), 201–230. https://www.sciencedirect.com/science/article/pii/S0272494405801847.

Wells, N., & Evans, G. (2003). A room with a view helps rural children deal with life's stresses. *Environment and Behavior, 35*(3), 311–330.

Wolf, K. L., & Flora, K. (2010). Mental health and function-A literature review. In *Green cities: Good health.* College of the Environment, University of Washington. www.greenhealth.washington.edu.

World Health Organization. (2016). *Urban green spaces and health.* Geneva: WHO Regional Office of Europe.

Bibliography

Amato, F., Pandolfi, M., Escrig, A., Querol, X., Alastuey, A., Pey, J., Perez, N., & Hopke, P. K. (2009). Quantifying road dust resuspension in urban environment by Multilinear Engine: A comparison with PMF2. *Atmospheric Environment*, *43*(17), 2770–2780.

Amato, F., Pandolfi, M., Viana, M., Querol, X., Alastuey, A., & Moreno, T. (2009). Spatial and chemical patterns of PM10 in road dust deposited in urban environment. *Institute of Earth Sciences* Jaume Almera, Spanish Research Council (CSIC), C/ Luis Solé Sabarís s/n, 08028 Barcelona, Spain.

American Chemical Society. (2009, July 23). Stop and smell the flowers – The scent really can soothe stress. *ScienceDaily*. www.sciencedaily.com/releases/2009/07/090722110901.htm. (retrieved 26 April 2023).

Belis, C. A., Peroni, E., & Thunis, P. (2013). Source apportionment of air pollution in the Danube region. http://iet.jrc.ec.europa.eu.

Biggs, C. (2019, May 10). Rooms with a view (and how much you'll pay for them). *The New York Times*. (retrieved 11 March 2023).

Cummings, B. E., & Waring, M. S. (2020). Potted plants do not improve indoor air quality: A review and analysis of reported VOC removal efficiencies. *Journal of Exposure Science and Environmental Epidemiology*, *30*(2), 253–261. https://doi.org/10.1038/s41370-019-0175-9.

Famulari, S. (2020). *Green up! Sustainable Design Solutions for Healthier Work and Living Environments*. New York: Productivity Press. https://doi.org/10.4324/978042297434.

Fernandez, M. (2006, October 29). A study links trucks' exhaust to Bronx schoolchildren's asthma. *The New York Times*, Region Section.

Gjording, L. R. (2022, October 25). Redlining and its impact on New York City. *City Signal*, Real Estate Section. (retrieved 11 March 2023).

Hale, T. (2022). Plants could be the key to a better semester. *Daily Universe*. (retrieved 31 March 2023).

Han, K.-T., & Ruan, L.-W. (2019). Effects of indoor plants on self-reported perceptions: A systemic review. *Sustainability*, *11*(16), 4506. http://doi.org/10.3390/su11164506.

Karagulian, F., Belis, C. A., Dora, C. F., Pruss-Ustan, A. M., Bonjour, S., Adair-Roahn, H., & Amann, M. (2015). Contributions to cities' ambient particulate matter (PM): A systematic review of local source contributions at global level. *Atmospheric Environment*, *120*, 475–483.

Maarten Hornikx, M. (2016). Ten questions concerning computational urban acoustics. *Building and Environment*, *106*, 409–421.

Nardone, A., Rudolph, K. E., Rachel Morello-Frosch, R., & Casey, J. A. (2021). Redlines and greenspace: The relationship between historical redlining and 2010 greenspace across the United States. *Environmental Health Perspectives*, *129*(1), 1. https://doi.org/10.1289/EHP7495.

Nuccitelli, D. (2023). *The little-known physical and mental health benefits of urban trees*. Yale Climate Connections. https://yaleclimateconnections.org/2023/02/the-little-known-physical-and-mental-health-benefits-of-urban-trees/. (retrieved 12 March 2023).

Pool, T. (2022). *On plant-associated microbiota in the indoors*. Green Architecture, Healthy Plants, Indoor Green. https://moss.amsterdam/2022/12/09/on-the-progressive-effects-of-plant-associated-microbiota-on-the-indoor-environment-2/. (retrieved 12 March 2023).

Raanaas, R. K., Horgen Evensen, K., Rich, D., Siostrom, G., & Patil, G. (2011). Benefits of indoor plants on capacity in an office setting. *Journal of Environmental Psychology*, *31*(1), 99–105.

Ulrich, R. S., Simons, R. F., Losito, B. D., Fiorito, L., Miles, M. A., & Zelson, M. (1991). Stress recovery during exposure to natural and urban environments. *Journal of Environmental Psychology*, *11*(3), 201–230. https://www.sciencedirect.com/science/article/pii/S0272494405801847.

Wells, N., & Evans, G. (2003). A room with a view helps rural children deal with life's stresses. *Environment and Behavior*, *35*(3), 311–330.

Wolf, K. L., & Flora, K. (2010). Mental health and function-A literature review. In *Green cities: Good health*. College of the Environment, University of Washington. www.greenhealth.washington.edu.

Wolverton, B. C., & Wolverton, J. D. (1993). Plants and soil microorganisms: Removal of formaldehyde, xylene and ammonia from indoor air environment. *Journal of Mississippi Academy of Sciences*, *38*(2), 11–15.

Wolverton, B. C. et al. (1989). *A study of interior landscape plants for indoor air pollution abatement: An interim report*. Mississippi: NASA.

World Health Organization. (2010). *WHO guidelines for indoor air quality: Selected pollutants*. Geneva: WHO Regional Office of Europe.

World Health Organization. (2016). *Urban green spaces and health*. Geneva: WHO Regional Office of Europe.

Image Credits

CHAPTER 2

2.1	Stevie Famulari Gds
2.2	Stevie Famulari Gds
2.3	Stevie Famulari Gds
2.4	Stevie Famulari Gds
2.5	Shutterstock
2.6	Shutterstock
2.7	Shutterstock
2.8	Shutterstock
2.9	Google Maps, Bronx, NY, retrieved March 2023
2.10	Stevie Famulari Gds
2.11	Shutterstock
2.12	Shutterstock
2.13	Stevie Famulari Gds
2.14	Shutterstock
2.15	Shutterstock
2.16	Stevie Famulari Gds
2.17	Stevie Famulari Gds
2.18	Stevie Famulari Gds
2.19	Stevie Famulari Gds
2.20	Stevie Famulari Gds
2.21	Shutterstock
2.22	Shutterstock
2.23	Trenton Moore
2.24	Stevie Famulari Gds
2.25	Shutterstock
2.26	Shutterstock
2.27	Stevie Famulari Gds
2.28	Shutterstock
2.29	Shutterstock
2.30	Shutterstock
2.31	Stevie Famulari Gds
2.32	Shutterstock
2.33	Shutterstock
2.34	Stevie Famulari Gds
2.35	Shutterstock
2.36	Shutterstock
2.37	Shutterstock
2.38	Stevie Famulari Gds
2.39	Shutterstock
2.40	Shutterstock
2.41	Stevie Famulari Gds
2.42	Shutterstock
2.43	Shutterstock

CHAPTER 3

3.1   Stevie Famulari Gds
3.2   Shutterstock
3.3   Stevie Famulari Gds
3.4   Stevie Famulari Gds
3.5   Shutterstock
3.6   Shutterstock
3.7   Shutterstock
3.8   Stevie Famulari Gds
3.9   Shutterstock
3.10   Stevie Famulari Gds
3.11   Stevie Famulari Gds
3.12   Stevie Famulari Gds
3.13   Stevie Famulari Gds
3.14   Shutterstock
3.15   Stevie Famulari Gds
3.16   Shutterstock
3.17   Stevie Famulari Gds
3.18   Stevie Famulari Gds
3.19   Stevie Famulari Gds
3.20   Stevie Famulari Gds
3.21   Shutterstock
3.22   Shutterstock
3.23   Shutterstock
3.24   Stevie Famulari Gds
3.25   Stevie Famulari Gds
3.26   Shutterstock
3.27   Shutterstock
3.28   Shutterstock
3.29   Shutterstock
3.30   Shutterstock
3.31   Stevie Famulari Gds
3.32   Stevie Famulari Gds
3.33   Shutterstock
3.34   Shutterstock
3.35   Shutterstock
3.36   Shutterstock
3.37   Shutterstock
3.38   Stevie Famulari Gds
3.39   Stevie Famulari Gds
3.40   Shutterstock
3.41   Shutterstock
3.42   Shutterstock
3.43   Shutterstock
3.44   Stevie Famulari Gds

CHAPTER 4

4.1 Stevie Famulari Gds
4.2 Stevie Famulari Gds
4.3 Stevie Famulari Gds
4.4 Stevie Famulari Gds
4.5 Shutterstock
4.6 Shutterstock
4.7 Shutterstock
4.8 Shutterstock
4.9 Shutterstock
4.10 Stevie Famulari Gds
4.11 Stevie Famulari Gds
4.12 Stevie Famulari Gds
4.13 Shutterstock
4.14 Shutterstock
4.15 Shutterstock
4.16 Shutterstock
4.17 Stevie Famulari Gds
4.18 Stevie Famulari Gds
4.19 Amber Leigh
4.20 Amber Leigh and Shutterstock
4.21 Shutterstock
4.22 Shutterstock
4.23 Trenton Moore
4.24 Stevie Famulari Gds
4.25 Stevie Famulari Gds
4.26 Shutterstock
4.27 Shutterstock
4.28 Shutterstock
4.29 Shutterstock
4.30 Shutterstock
4.31 Stevie Famulari Gds
4.32 Stevie Famulari Gds
4.33 Shutterstock
4.34 Shutterstock and Stevie Famulari Gds
4.35 Shutterstock
4.36 Shutterstock
4.37 Shutterstock
4.38 Stevie Famulari Gds
4.39 Stevie Famulari Gds
4.40 Stevie Famulari Gds
4.41 Stevie Famulari Gds
4.42 Shutterstock
4.43 Shutterstock
4.44 Stevie Famulari Gds

CHAPTER 5

5.1   Stevie Famulari Gds
5.2   Shutterstock
5.3   Stevie Famulari Gds
5.4   Shutterstock
5.5   Shutterstock
5.6   Stevie Famulari Gds
5.7   Shutterstock
5.8   Shutterstock
5.9   Shutterstock
5.10   Stevie Famulari Gds
5.11   Shutterstock
5.12   Shutterstock
5.13   Stevie Famulari Gds
5.14   Shutterstock
5.15   Shutterstock
5.16   Shutterstock
5.17   Stevie Famulari Gds
5.18   Stevie Famulari Gds
5.19   Shutterstock
5.20   Shutterstock
5.21   Shutterstock
5.22   Shutterstock
5.23   Shutterstock
5.24   Stevie Famulari Gds
5.25   Shutterstock
5.26   Shutterstock
5.27   Stevie Famulari Gds
5.28   Shutterstock
5.29   Shutterstock
5.30   Shutterstock
5.31   Stevie Famulari Gds
5.32   Shutterstock
5.33   Shutterstock
5.34   Stevie Famulari Gds
5.35   Shutterstock
5.36   Shutterstock
5.37   Stevie Famulari Gds
5.38   Stevie Famulari Gds

Index